Renewable Energy Conversion Systems

Renewable Energy Conversion Systems

Muhammad Kamran
Department of Electrical Engineering and Technology,
Riphah International University, Pakistan

Muhammad Rayyan Fazal
Department of Electrical Engineering and Technology,
Riphah International University, Pakistan

ELSEVIER

ACADEMIC PRESS
An imprint of Elsevier

Academic Press is an imprint of Elsevier
125 London Wall, London EC2Y 5AS, United Kingdom
525 B Street, Suite 1650, San Diego, CA 92101, United States
50 Hampshire Street, 5th Floor, Cambridge, MA 02139, United States
The Boulevard, Langford Lane, Kidlington, Oxford OX5 1GB, United Kingdom

British Library Cataloguing-in-Publication Data
A catalogue record for this book is available from the British Library

Library of Congress Cataloging-in-Publication Data
A catalog record for this book is available from the Library of Congress

ISBN: 978-0-12-823538-6

For Information on all Academic Press publications
visit our website at https://www.elsevier.com/books-and-journals

Publisher: Joe Hayton
Acquisitions Editor: Lisa Reading
Editorial Project Manager: Sara Valentino
Production Project Manager: Poulouse Joseph
Cover Designer: Vicky Pearson

Typeset by MPS Limited, Chennai, India

Dedication

To my parents Muhammad Ramzan (late) and Naziran BiBi for their support and love and to my beloved wife Badar Un Nisa and my son M. Sarim Kamran for their inspiration and love.

Muhammad Kamran

To my family, for their continuing unconditional love and support throughout my life. I am also thankful to my teachers who equipped with me the power of knowledge. Last but not least, I am grateful to Mr. Muhammad Kamran, who provided me with this great opportunity to work with him.

Muhammad Rayyan Fazal

Contents

Chapter 1

Fundamentals of renewable energy systems

Muhammad Kamran and Muhammad Rayyan Fazal
Department of Electrical Engineering and Technology, Riphah International University, Pakistan

1.1 Introduction

1.1.1 Why renewables

Until the 19th century, the main concern of humans was to gather food for family members, and technological progress, as a result, remained limited. Energy and fuels played a vital role in the evolution of human civilization since humans learned to ignite the fire for the very first time. The primary energy source was wood until 1700 and then coal superseded it in the 1780s. A rapid transition can be seen in the past three centuries including steam engines, enhancement in oil extraction and refinement techniques, and coal-based power plants that dramatically altered this fuel usage pattern. Hence, it took just two centuries to shift towards fulfilling 80% of energy requirements through fossil fuels whereas it took thousands of years to switch from wood to coal. Until now, coal is playing a vital role in fulfilling energy demands. However, its use remained limited, i.e., electricity production, till the development of the techniques to convert it into chemicals [1]. Their practical implementations are yet to be validated for commercial use before considering it as a replacement for fossil fuels. Coal consumption is further reduced because of the recent developments in natural gas, and methane gas [2]. On the other hand, there are many types of fuels obtained through fossil fuels such as liquid gas, crude oil, gasoline, kerosene, diesel, etc. A large amount of chemicals is produced through crude oil. Furthermore, there is no additional requirement of chemical processes in using them as fuels hence byproducts are usually of a negligible amount. That's why crude oil remained the central point for technology development and policymaking in the last century.

Today, the dominating power in the world is technology. Hence, the demand to power various devices required for smooth and efficient operations

Renewable Energy Conversion Systems. DOI: https://doi.org/10.1016/B978-0-12-823538-6.00009-9

has been dramatically increased. The technology evolution and rapid population growth are pushing the countries to look for various sustainable alternative forms of energy to accommodate future needs. With industrialization and the increasing population, energy demands have been increased exponentially whereas the conventional energy sources are depleting at the same rate. As per estimations, 80% of the world's energy is taken from non-renewable resources i.e., oil, coal, and natural gas. These fossil fuels are of limited quantity and likely to get exhausted soon if this trend continues. As fossil fuels are unlikely to meet the demand of the exponentially growing population of the world. In the same way, the emissions like CO_2 are reaching dangerous thresholds due to excessive use of fossil fuels. This is now becoming a serious concern because of the increasing threat to the overall ecosystem. It is not only the greenhouse gases that are causing the problems but these fuels also left Sulphur, nitrogen, and carbon oxides that significantly degrade the environment. However, there are now rules and regulations strictly implemented by most of the countries to narrow down the potential hazards at large but yet the amount of fossil fuel usage in various applications around the globe has risen so high that it seems impossible to get complete rid of the pollutants. Another associated problem with fossil fuels is their non-uniform distribution worldwide. There exists a huge gap in the demand and supply of fossil fuels throughout the world. It makes some nations dependent upon others, and any sociopolitical conflicts or economic instabilities might result in massive energy-related problems. Therefore, there is a sheer need of getting rid of this huge dependency and look for alternative technologies superior in environmental factors as well as sustainable point of view.

This worldwide crisis can be best addressed utilizing 'renewable energy resources', referred to as systems having the ability to provide unlimited energy potential. There are plenty of these types available these days with their technologies matured and accepted worldwide. Research on reliability, efficiency, and consistency is still underway but yet these technologies are now widely implemented to produce energy. These renewable energy sources can be used to generate electricity or to meet the heating requirements with minor or low impact on the environment. Renewable energy resources like Solar (Photovoltaic Systems), Wind (Wind turbines), the heat of the earth core (Geothermal Energy), biomass (Bioenergy), and water (hydropower energy) have emerged as strong competitors of fossil fuels. Unlike fossil fuels, many renewables are unable to be used directly for energy production or producing chemical byproducts. Renewable resources such as solar, wind, and hydropower can act as sustainable sources and capable of fulfilling energy demands in different shapes. However, they cannot be used to generate any other form of fuel or chemicals. Still, they produce a sustainable form of energy as these resources are unlikely to be discontinued or politically affected. Sun will continue to shine for any reason and other resources are mainly dependent upon solar energy for their existence such as wind and water. Therefore, using these technologies for energy production will not affect any kind of resources, unlike fossil fuel reservoirs that are likely to get exhausted for our future generations. There are many advantages of using these renewable energy resources as energy-producing alternatives:

- ***Air Pollution:*** Due to the increasing demand for fossil fuels, their transportation, industrial usage, and power production is polluting the air to a great extent. The use of charcoal and wood is contributing to poor air conditions and also causing millions of premature deaths, according to the world health organization. Instead of polluting the environment, renewables are also saving precious resources hence better for our health.
- ***Greenhouse Gas Emission:*** Renewable Energy has a great advantage in achieving better climatic conditions over non-renewable energy. As the emissions from fossil fuels have a devastating effect on the environment. The combustion of these fossil fuels adds to the greenhouse gas emissions hence promoting the global warming phenomenon. Renewables on the other hand have little to no emission at all even if the full life cycle of the project is considered.
- ***Economics:*** Renewable energy is never subjected to get affected because of the geopolitical crisis, or discontinuity of supply, or sudden price spikes found to be common in the case of fossil fuels.
- ***Community's betterment:*** Generating power from renewable resources is an emerging technology. Many countries are putting huge investments in the research and development sector for their practical implementation. This renewable energy in return generates revenue and is used for further development of this sector, hence proving a better alternative for the community's betterment.
- ***Resilient infrastructure:*** Renewable energy facilitates the development of the energy production setup at remote locations. Hence, making urban independent from national grids. Power systems are becoming more flexible, manageable, and resilient against catastrophic events with the help of renewable energy units.
- ***Easy access:*** Technology evolution is leading towards cheaper power generation through renewable resources. This in return is making it possible to use renewable resources for energy production especially for the people living in remote locations.
- ***Secure and stable:*** Greater energy demands and modern infrastructure requires the adoption of safe, secure, and stable strategies for energy production. Renewable resources are safe and secure hence offering more stability in return.

There is a wide range of applications using renewable energy in different forms. Worldwide, 26% of the electricity generation is through renewables but yet electricity represents just 17% of the world's energy needs. A major form of energy usage is cooling and heating that comprised nearly 50% of the total world's energy demand. Whereas nearly one-third of the total energy is required by the transportation sector. The transportation and heating/cooling sector are utilizing less than 10% hence, bringing no help to decarbonization.

These renewable energy resources can be best utilized in various forms as there is still a lot of untapped potentials. They are used these days in building as solar thermal water heaters, and bio energy-based boilers. Direct heat from geothermal resources and heat pumps are also utilized in buildings. However, there is a lot of work yet to be required to fulfill the energy demand of the buildings through renewable energy resources. There is a massive room from the industrial point of view to use renewable energy in different manufacturing processes. The heat required for the food industry and paper industry can be taken from renewable resources. Renewable electricity is capable of producing enough hydrogen required for various industrial processes such as required in steel, iron, and cement industry. Reduction in industrial energy demand will help in decreasing their reliance upon fossil fuels to replace them with renewable options. Whereas, sustainable biofuels can be used in the transportation sector as a renewable energy option. The electric vehicle industry is now well established throughout the world. Renewable energy can be used to power these electric vehicles. The batteries in electric vehicles can be used to store power and deliver upon requirement. Renewables can be used to generate electro-fuels like hydrogen to be used as fuel in the aviation industry and shipping sector. However, there is a sheer need of reducing the energy demand in the transportation sector through formulating policies promoting energy conservation and efficiency. A cumulative strategy design for renewable energy and its efficient use is needed.

This book is designed to examine these renewable energy options in detail including an investigation of their associated technologies, design, modeling, and their applications with current and future research trends. Also, this book provides a detailed overview of power electronics used for the integration of these systems with power systems and thermodynamic laws.

1.1.2 Types of energy

In nature, energy is found in different forms and is responsible to perform different processes. The transformation of energy from one form to another form is according to the law of conservation of energy or the first law of thermodynamics that states "energy cannot be created nor destroyed but can be transformed from one form to another". One form of energy is the kinetic energy that develops due to the motion of the object. The energy in the wind and the energy in the moving water is the kinetic energy. The other form of energy is the potential energy that exists in the objects or the molecules due to their position with respect to earth. Energy stored in falling water, energy in the compressed spring, and the energy in the hanging fruit before fall are examples of potential energy. Different forms of energy are electrical energy, mechanical energy, chemical energy, wave energy, tidal energy, thermal energy, magnetic energy, nuclear energy, geothermal energy, and geothermal energy. All these types of energy are interchangeable. The burning of fossil fuels, biomass, and biodiesels to generate electricity contains three energy conversions. The chemical energy of the fuel is converted into thermal energy in the boiler, thermal energy is converted into mechanical energy of the turbine, mechanical energy of the turbine is converted into electrical energy in the electrical generator.

1.1.3 Conventional and renewable energy

Conventional energy: The energy sources that once exhausted, do not replenish themselves within a specific period are called conventional or non-renewable energy sources like coal, gas, and oil. For a long time, these energy sources have been used extensively to meet the energy demands. As the rate of consumption is much greater than the rate of formation, these sources of energy have been depleted and do not replenish. Conventional sources of energy emit hazardous emissions that not only damage the earth's atmosphere also deteriorate the health conditions of the livings.

Renewable energy: Those sources of energy that are replenished by a natural process after being used are renewable energy sources. These sources are solar energy, wind energy, geothermal energy, bioenergy (biomass, biogas, and biodiesel), geothermal energy, and hydro energy. Almost all these renewable energy sources exist because of the sun. Solar PV uses the solar irradiance to kick out the electrons from the valance band to the conduction band that constitutes the electric current. Wind energy exists owing to the bumpy heating of the earth's surface causing density variation. Owing to the sunlight, water evaporates from the earth's surface and causes clouds in the sky that completes the life cycle of the water causing the existence of hydro energy in the high potential areas. Different renewable energy technologies and their end products are given in Table 1.1

1.1.4 SWOT analysis of the renewable energy

1.1.4.1 Strength

- Renewable energy sources are abundantly and freely available in almost all parts of the world.
- Validated and authenticated global atlas for renewable energy sources.
- Private investors have a keen interest in investing in renewable energy sources owing to lucrative incentives from the government.
- Renewable energy sources have minimum emissions as compared to conventional energy sources. However, the production of equipment used in renewable energy technologies might have a few emissions.
- Minimum operation and maintenance cost
- Highly public acceptance

TABLE 1.1 Renewable energy technologies and their end products.

Technology	End Product
	Solar Energy
• Solar photovoltaic • Solar concentrated power • Passive solar energy • Low-temperature solar energy	DC electricity Heat, steam, electricity Heat, light, cold, ventilation Cooking, drying, space heating, and water heating
	Wind Energy
• Offshore wind turbines • Onshore wind turbines • Small wind machines	Electricity Electricity Electricity, Movement
	Hydro Power
• Mini/Micro-hydro • Small and large hydropower	Movement, electricity Electricity
	Bioenergy
• Combustion • Gasification • Hydrolysis-Fermentation • Anaerobic digestion • Pyrolysis • Transesterification • Palletization	Heat, steam, electricity, CHP Heat, electricity, CHP Ethanol Biogas Biofuels Biodiesel Pellets, heat, electricity
	Geothermal Energy
• Heat pumps • Direct heating • Power production	Heat Heat, steam Electricity
	Ocean Energy
• Tidal energy • Wave energy	Electricity Electricity

1.1.4.2 Weakness

- The capital cost of the renewable energy systems is quite high as compared to the running cost.
- Wind turbines and hydro turbines cause the noise pollution.
- The renewable energy sources are intermittent and are not available 24 hours. So we have to use extra equipment like inverters, rectifiers, and storage systems that increases the system cost.
- Farmers are ignorant of the energy crops.

- The construction of large dams requires a long period.
- Sedimentations in the dams deteriorate the storage capacity.

1.1.4.3 Opportunities

- Renewable energy technologies are developing progressively with the betterment of efficiency.
- The growing energy demand and the depleting fossil fuels allow the flourishment of renewable energy.
- Renewable energy systems can be installed at a small level called off-grid systems. The areas far from the grid can be electrified by the locally available renewable energy sources.
- Renewable energy systems are socially and economically accepted by the people.
- The construction of large dams requires a long period.
- By investing in R&D, renewable energy technologies, and a balance of systems can be made cheap that will help achieve the grid parity.

1.1.4.4 Threats

- Conventional energy technologies are more matured than renewable energy technologies that could hinder the adoption of renewable energy sources.
- In some countries, there are no incentives to the renewable energy systems that could also hinder the public interest in investing in renewable energy.
- With the growing energy demand, energy security needs to be ensured and the renewable energy economy needs to be more sustainable.

1.1.5 Global warming and climate change

Though Global warming and climate change are most often interchangeably used, both of them have dissimilar meanings. During industrialization, human activities of burning fossil fuels to meet the energy demands have increased the earth's temperature. Since the pre-industrialization (1850–1900), the earth's average temperature has been increased by 1°C because of human activities. The current increment in the earth's temperature is 0.2°C/decade.

Carbon dioxide emissions from renewable energy sources are minimum as compare to conventional fossil fuels. Lignite has a major contribution toward the CO_2 emissions that is 1054 g CO_2-eq/kWh. Similarly, other conventional energy sources have a considerable contribution in CO_2 addition to the atmosphere. The detail of CO_2 emission from each conventional energy source is shown in Fig. 1.1. On contrary to conventional energy sources, renewable energy sources have little emissions of CO_2. Offshore wind energy has only 9 g CO_2-eq/kWh. The detail of CO_2 emission from renewable energy sources is also shown in Fig. 1.1.

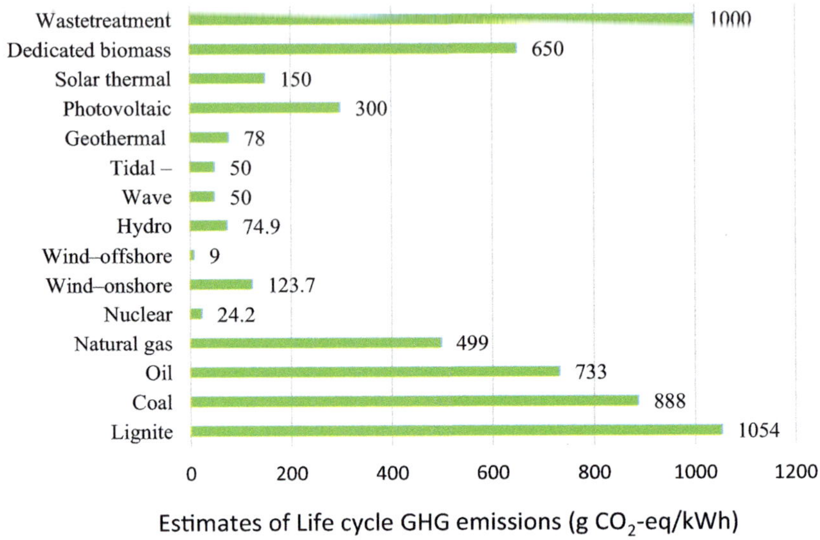

FIGURE 1.1 Estimates of Life cycle GHG emissions (g CO_2-eq/kWh).

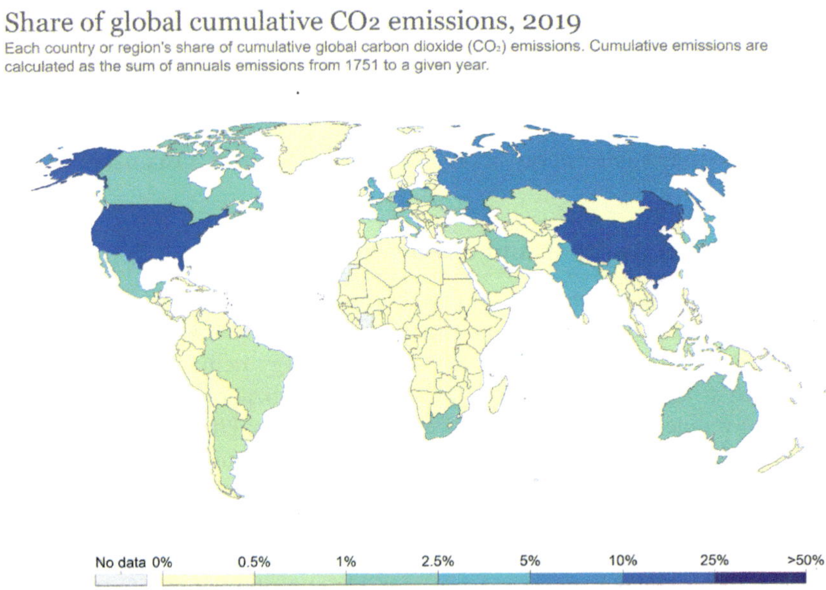

FIGURE 1.2 Global share to CO_2 emissions [3].

Climatic change has now been considered the main concern of this century. Whereas the global share of CO_2 emissions until 2019 has been shown in Fig. 1.2. United States leading the figures with 399 Billion tones emission of CO_2 and China is the second-largest contributor with half of the total emissions as compared to the US. Whereas most of the countries of Africa and Pakistan from Asia contributed less than 0.01% of the total emission for more than the last 250 years [3] (Fig. 1.3).

Countries have agreed in Paris agreement to keep temperature rise well below 2°C and ideally to 1.5°C. this Paris agreement is a deal within United Nations Framework Convention on Climate Change (UNFCCC) related to a reduction in climatic change signed in the year 2016. Hence to achieve this ambitious target, CO_2 emissions have to be reduced nearly 3.5% on annual basis continuously till 2050 and even after that. Renewable energy generation along with electrified transportation and heat is expected to contribute 60% in CO_2 emission reduction required by 2050. This contribution is estimated to be 75% if direct use of renewables is also included and reaches even 90% upon adding energy efficiency as shown in Fig. 1.4 [4].

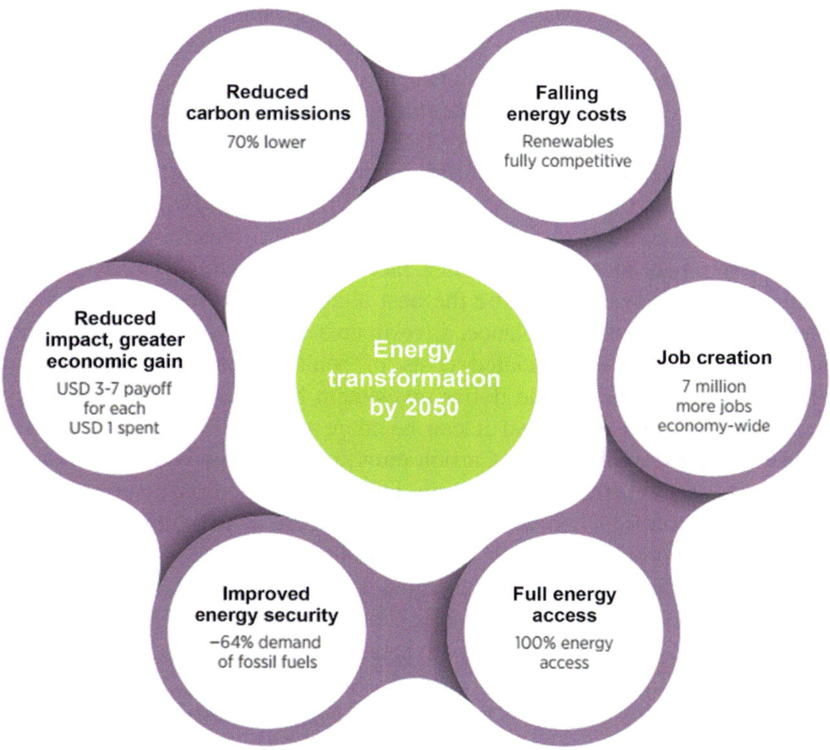

FIGURE 1.3 Drivers for world energy transformation [4].

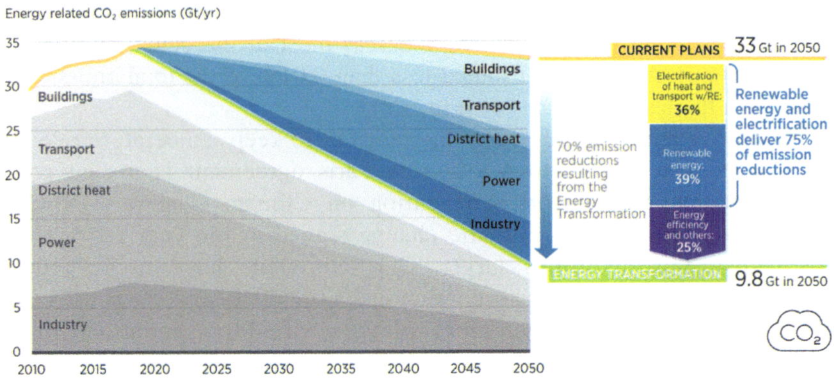

FIGURE 1.4 Efficient energy usage along with renewables can bring down the 90% of CO_2 emissions by 2050 [4].

1.1.6 World energy transformation by 2050

International Renewable Energy Agency (IRENA) has devised a roadmap for renewable energy depicting the renewable-based energy contribution by 2050 worldwide [4]. Though IRENA has declared the reduced CO_2 emission to be the main driving force of this transformation, however, many other factors are contributing towards this change too, as shown in Fig. 1.4. The main factor involved is related to the declining cost of renewable energy. As for onshore wind turbine projects in 2018, USD 0.056 cost per kilowatt-hour (kWh) was observed that is 13% less than as compared to that of 2017 and 35% lower than the cost in 2010. Whereas solar energy has seen a 77% decline in cost in 2018 as compared to 2010 [4]. And this trend is very likely to be continued in future. Another factor is air quality improvement, as renewables are considered to be the best alternatives to fossil fuel energy as they are clean by nature. Hence, investments in renewable energy systems will bring down the associated costs of health issues and environmental effects. Against every single dollar investment by 2050, the expected payoff is a minimum of USD 3 and it can be more than USD 7 relying upon the usage of the externalities [4]. Carbon emission reduction is another factor to achieve by 2050, 70% reduced emissions are targeted as compared to the present conditions. This worldwide transformation will also enhance the energy security, accessibility, and affordability of nations. Countries heavily relying upon fossil fuels are subjected to security issues. Investments in renewable energy systems will make them independent of this dependency hence, offering flexibility and resilience. Also, it serves for improving the life quality of humans as it can provide electrification in rural areas where the national grid is unable to be extended. More interestingly, these renewable energy resources also provide socio-economic benefits which play a vital role against any political decisions. The renewable industry creates jobs to engage men and women of various backgrounds. The investment in this sector brings technological advancements, attracting grants for relevant projects. While countries lagging will have to just switch from importing fossil fuels to renewable energy-related equipment [4].

Another way to decarbonize the energy systems is by increasing electricity production from renewable sources. The integration of electrified transportation and heat along with economical renewable electricity production will serve as a healthy combo for energy transformation. IRENA's renewable energy roadmap (REmap) has set the threshold of 86% of the renewable energy share in electricity generation by 2050. Whereas the electricity share of energy consumption will rise from 20% today to 50% by 2050. The energy consumption by industrial and domestic users is almost estimated to get doubled and transportation will require 40% by 2050 as compared to 1% of today [4]. Wind and solar are undoubtedly global leaders in this transformation. Wind is expected to produce one-third of the world's total energy by 2050. Whereas, if we look for the overall installed capacity, the solar industry would require much larger expansion due to the lower capacity factors attained by solar-related projects [4] (Fig. 1.5).

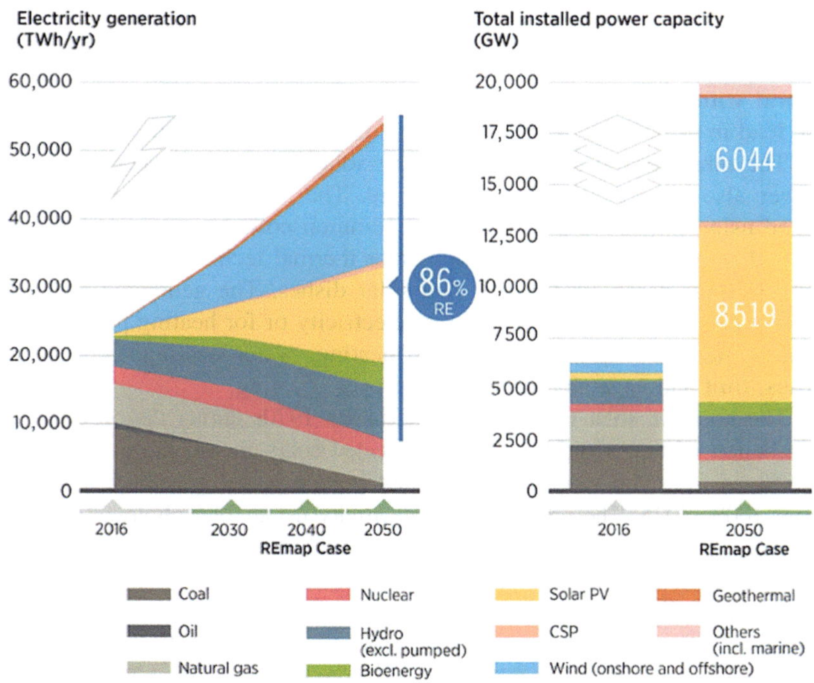

FIGURE 1.5 Electricity generation and installed capacity, the scenario of Energy Mix, present and future.

1.1.7 Prospects of renewable energy in the world

1.1.7.1 Solar energy

The sun is an everlasting and a giant source of energy. Almost all renewable energy sources are directly or indirectly completing their life cycle because of solar energy. The direct utilization of solar energy is solar photovoltaic and solar thermal. In solar photovoltaic technology, light energy termed as photon emits the electrons from the valance band of the semiconductor material. If the energy of the photon is larger than the bandgap energy of the semiconductor material, the electron jumps from the valance band to the conduction band and constitutes the current through the photovoltaic material. If the energy of the photon is smaller than the bandgap energy of the semiconductor material, the electrons do leave the valance band but before entering the conduction band they lose all their energy and come back to the valance band. The solar photovoltaic system could be off-grid or grid-connected. The balance of the system of the solar photovoltaic system includes a mounting system, DC-DC converter, Maximum Power Point Tracker (MPPT), inverter, DC cables, and battery bank. As solar energy is intermittent and is not capable of providing energy constantly so for continuous operation of the standalone solar PV system, a battery backup is required that increases the cost of the system. On the other hand, the grid-connected solar PV system can be used in standalone mode or the grid integration model. To overcome the intermittency of solar energy, solar PV is used in hybrid with other renewable energy sources and a diesel generator to meet the load in peak hours when solar PV is unable to meet all the load.

The other technology for harnessing solar energy is concentrating solar power also termed solar thermal energy. The solar radiations are concentrated using mirrors and focused on a common collector where solar energy is used to heat the water or the air. Solar thermal technologies include solar towers, solar parabolic troughs, and solar dishes. The generated steam and the hot air are used to either generate electricity or for heating purposes. The heating purposes may include space heating, water heating, cooking purposes, fruit and vegetable drying, and process heating.

The boom in solar technology came in the 1970s during the oil crisis. The installation of the off-grid and grid-connected solar photovoltaics is increasing every year. Fig. 1.6 shows the solar photovoltaic power generation every year from 2000–2025. The production of electricity using solar photovoltaic is rapidly increasing in a way that most of the countries have achieved grid parity. Fig. 1.6 shows that the generation of electricity using solar PV in 2018 was 554 TWh that is expected to increase to 1600 TWh in 2025. Similarly, the generation of electricity through concentrating solar power in 2018 was 11.3 TWh which is likely to reach 26 TWh in 2025. From Figs. 1.6 and 1.7 it can be estimated that the capacity addition and the cumulative capacity of solar energy rapidly increased in the year 2010–2011 [5].

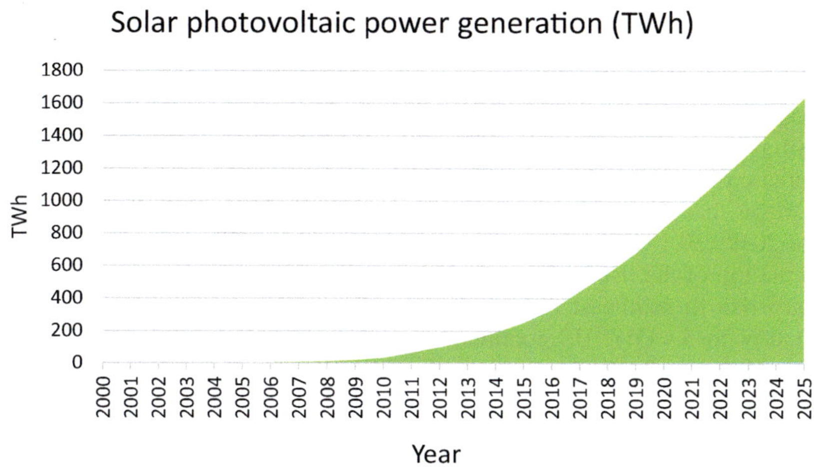

FIGURE 1.6 Global net addition in solar photovoltaic power generation.

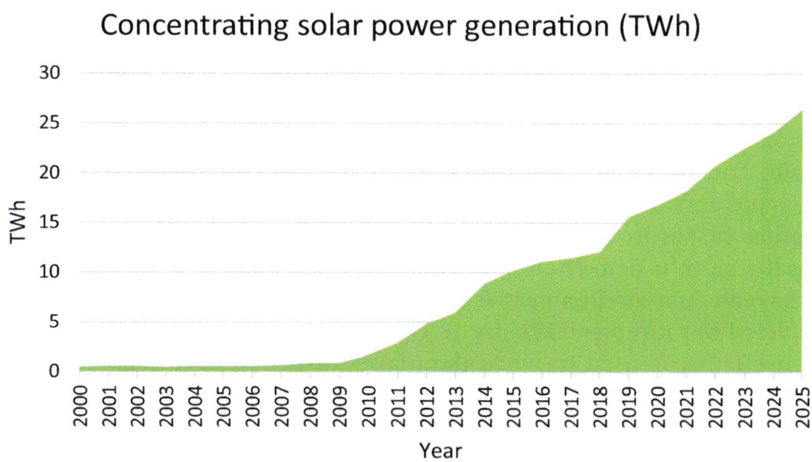

FIGURE 1.7 Global net addition in concentrating solar power generation.

Solar energy observed a sharp rise globally in 2019 with the net addition of 22% (131 TWh) taking the total figure to 720 TWh. Solar is sharing 3% of the overall global electricity generation. it superseded bioenergy to become the third-largest power generation source followed by hydropower and onshore wind. It is the second-largest growth among all sorts of renewable energy generation in 2019 with wind energy leading the figures and hydropower slightly behind solar. Solar energy saw a record-breaking year in terms of its capacity addition and the trend is likely to be continued.

The average growth of solar energy is expected to be 15% annually taking the solar energy from 720 TWh in 2019 to almost 3300 TWh in 2030 [6].

The growth of solar energy addition remained exponential over the last decade. Government policies and strong support motivated the industrial as well as domestic users to add solar energy into their systems. Solar capacity increased in the US, Europe, Middle East, Africa, and Latin America. China, however, has seen some decline due to revised policies in 2018 (44.4 GWh) and 2019 (30.1 GWh) as compared to 2017 (53.1 GWh) but yet it remained on the top of the list as shown in Fig. 1.6. PV generation has also observed a sharp rise in southeast Asia with only Vietnam's capacity increasing from 0.1 GW to 5.4 GW. US continued to follow a stable growth with 13.2 GW net addition in 2019. Spain, Netherland, and Germany help European Union to observe 98% capacity for the same year. In Brazil, record-breaking 2.1 GW solar were installed in a single year due to the introduction of generous incentives and consumer-friendly policies by the government. However, the growth is expected to accelerate due to its cost getting cheaper as compared to its competitors [7].

1.1.7.2 Wind energy

Another renewable energy source is wind energy whose primary reason is the sun. Because of the uneven heating of the earth's surface and the air on the earth, the density of the air changes at various locations on the earth. The air on the earth region where the sun radiations heat up the air becomes less dense and moves upward. Whereas the cool air from other regions rushes to fill that gap. About 1 to 3 % of the sun's energy reaching the earth is converted into wind energy. Wind energy is considered a renewable and environment-friendly source of energy as the air is never depleted but it replenishes itself. The moving air contains kinetic energy in itself that is used to move the blades of the wind turbine converting kinetic energy into mechanical energy. This mechanical energy is converted into electrical energy through a generator coupled with a wind turbine.

The future of wind energy is very bright, this is mainly because of the decline in fossil fuel reserves, huge cost associated with fossil fuel exploration, and shut down of coal power plants due to environmental hazards. Developing countries can take maximum benefit out of it, especially where energy poverty is a big challenge. The wind turbines can be economically viable in rural areas and remotely located areas having no access to grid electricity. It is a very fast-growing market that is getting mature. Developing countries invested heavily in the wind to cater to environmental problems but now it is economically well justified too.

According to the International Energy Agency (IEA), the total electricity production by wind energy in 2018 was 1273 TWh which is expected to increase to 2500 TWh in 2025. Fig. 1.8 shows the global net addition in the wind power generation every year.

According to the International Energy Agency (IEA), onshore wind energy increased 12% in 2019 whereas offshore increased by 20% in 2018, a 12% decline in growth as compared to 2017 (32%). Europe remained the leading region as far as technological advancements, rapid growth, and deployment is concerned. The onshore wind technology remained the largest capacity renewable power (non-hydro), with generations almost equal to all others combined. IEA reports that to achieve the goals set by the Sustainable development scenario (SDS) 10% annual growth is required till 2030. The net onshore wind power generation is 1323 TWh in 2019 which is expected to be 3749 TWh [8].

China remained the leading country in 2019 with 23.8 GW also in 2019 as compared to 19 GW of 2018 and 14.5 GW of 2017. European union added 9.1 GW in 2019 that is 17% more as compared to 2018 with Spain, Sweden, and France being the top three leading countries. Whereas the United States added 9.1 GW as compared to 6.9 GW of 2018 [8].

The current installed power generation capacity of offshore wind turbines is 67 TWh which is expected to be 3.8 TWh in 2025 and 606 TWh by 2030 as shown in Fig. 1.9. The grid-connected installed capacity of an off-shore wind turbine is 5.9 GW in 2019, 40% high as compared to that in 2018. China and European Union have observed a sharp rise in the installation. China leading the figures with 2.3 GW capacity additions in 2019 with United Kingdom (1.6 GW) and Germany (1.1 GW) being second and third in the list. Nevertheless, accelerated growth is required to achieve the targets by overcoming the barriers. The national policies will remain critical for its further development and accelerated growth [9].

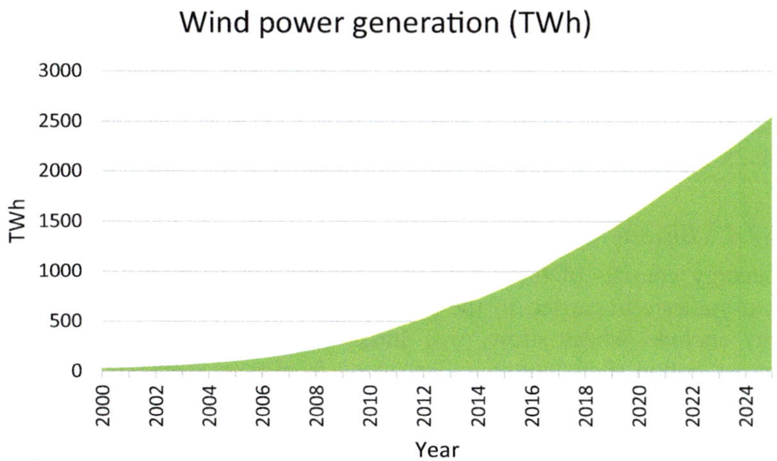

FIGURE 1.8 Global net addition in wind power generation.

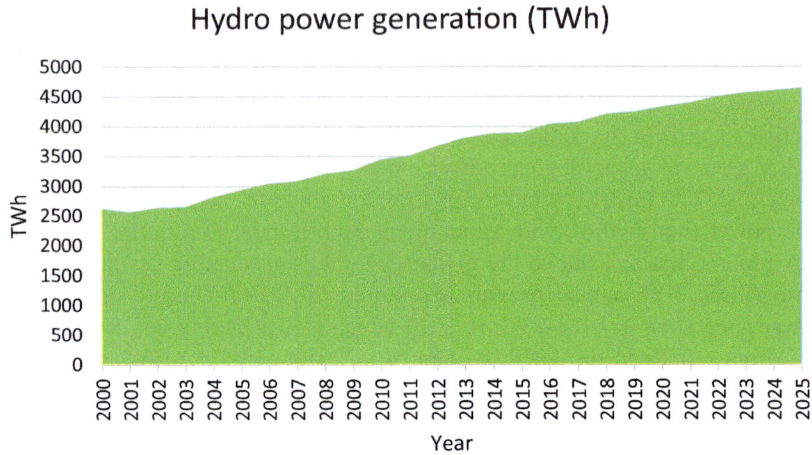

FIGURE 1.9 Global net addition in hydropower generation.

1.1.7.3 Hydropower

The life cycle of the water is also owing to the sun starting from the melting of the glaciers. During summer, when the sun shines for more than 8 hours a day, glaciers melt and the water reaches the river, canals, and then fields. In the course of water where the level of the water channel is higher than the earth level, the potential energy of the water is harnessed using a hydro turbine coupled to an electrical generator. Among all the renewable energy sources and corresponding technologies, hydropower is the most mature one and the cost of energy from hydropower is the least among other conventional and non-conventional energy sources because it does not use polluting fuel to operate. It means that hydropower does not emit carbon dioxide in the atmosphere making it a green energy source.

The generation of hydroelectricity is larger than any other renewable and conventional energy source. From Fig. 1.9 it can be seen that in 2000 the generation of electricity through hydropower was 2500 TWh which increased to 4325 TWh in 2018 and it is expected to increase to 4600 TWh in 2025.

1.1.7.4 Bioenergy

Bioenergy consists of the energy from plants and animals. The process of photosynthesis is carried in the presence of sunlight. The sources of bioenergy include wheat straw, rice husk, sugarcane tops, sugarcane trash, bagasse, cotton shells cotton stalks, wood from trees, herbaceous trees, animal dung, animal manure, and solid waste. The biomass can be used as an alternative to coal for coal-fired power generation that can further be used either for heating purposes or the generation of electricity. Whereas various other biomass sources can be used to generate biogas by the anaerobic

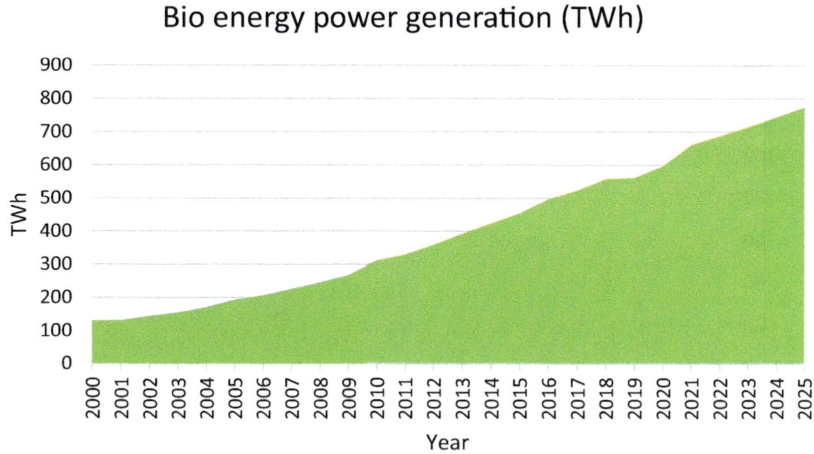

FIGURE 1.10 Global net addition in bioenergy power generation.

digestion process. Some of the biomass can be used to generate chemicals and biodiesel.

In rural areas the energy needs are fulfilled by burning the woody plants collected from the fields, cotton shells and cotton stalks, rice husk and sugar cane trash. The burning of biomass to get heat energy is an old concept as human history on the earth. The modern briquetting and palletizing techniques have made it easier to handle and transport the biomass. These techniques have also made the biomass dense energy source to be fed to the power plants.

The IEA has shown in Fig. 1.10 that in 2000 only 100 TWh electricity was being produced by the bioenergy sources that increased to 550 TWh in 2018 and expected to reach 7900 TWh in 2025. In 2018, 8.349 TWh electricity was being produced by liquid biofuels, 88.986 TWh by biogas, and 421.131 TWh by the primary solid biofuels as shown in Fig. 1.12.

1.1.7.5 Geothermal

The energy contained by the fluids and the rocks beneath the earth's surface is called geothermal energy. The energy of the earth is harnessed by digging wells. The water is sent through the well and steam and hot water are collected out of the well. The steam collected is used to drive a turbine coupled with the electrical generator. The steam and hot water can also be used for heating and cooling purposes depending on the temperature of the steam and hot water. Geothermal power plants are categorized as flash steam, dry steam, and binary. In flash steam power plants high-pressure hot water is drained out of the earth and in a flash chamber converted into steam to operate a steam turbine coupled to the electrical generator. After passing through

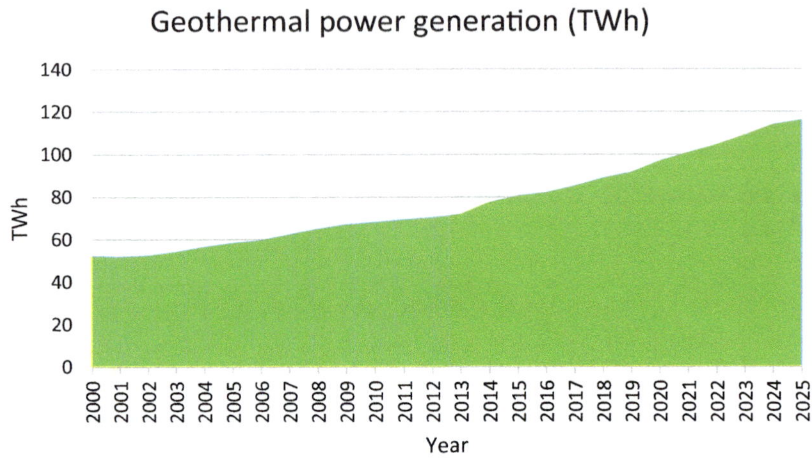

FIGURE 1.11 Global net addition in geothermal power generation.

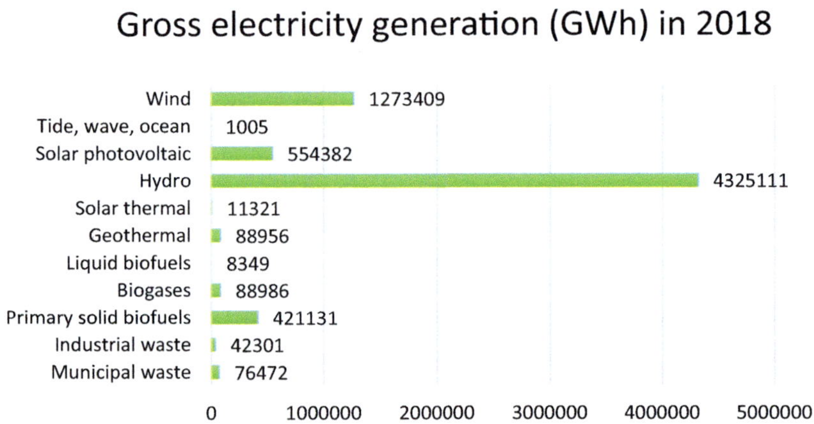

FIGURE 1.12 Gross Electricity Generation (GWh) in 2018.

the turbine, steam is condensed and sent back to the earth to reuse. Most of the geothermal power plants use flash steam technology to generate electricity. Dry steam is the oldest method to tap geothermal energy by taking steam coming out of the fractures on the earth that is used to drive the turbine to generate electricity. In binary geothermal power plants, the hot water or the steam never comes in contact with the steam turbine because of the application of the heat exchanger. Binary power plants use a secondary (binary) working fluid with a boiling point much lower than the water that takes the heat from the primary fluid through the heat exchanger. In these power plants, no material comes in contact with the atmosphere making it environment friendly.

According to IEA in 2000, 52.2 TWh electricity was being produced by the geothermal power plants that raised to 88.956 TWh in 2018 and it is expected to reach 116.3 TWh in 2025. Fig. 1.11 shows the global net addition of electricity generation by geothermal from 2000–2025.

1.1.8 The structure of the book

The book consists of nine chapters: introduction, thermodynamics of renewable energy systems, power electronics for renewable energy systems, solar energy, wind energy, hydro energy, fuel cell, bioenergy, and geothermal energy. This book goes beyond the potential assessment, theoretical aspects of advances in renewable energy, and addresses future trends. By focusing on the design of developing technologies, relevant operation and detailed background, and an understanding of the application of power electronics and thermodynamics processes in renewable energy, this book provides an analysis of advancing energy systems. The book will be of interest to engineering graduates, researchers, professors, and industry professionals involved in the renewable energy sector and is ideal for advanced engineering courses dealing with renewable energy, sources, thermal and electrical energy production, and sustainability.

References

[1] H.H. Schobert, C. Song, Chemicals and materials from coal in the 21st century, Fuel vol. 81 (1) (2002) 15–32. 2002/01/01/.

[2] S.H. Mohr, J. Wang, G. Ellem, J. Ward, D. Giurco, Projection of world fossil fuels by country, Fuel vol. 141 (2015) 120–135. 2015/02/01/.

[3] H. Ritchie, Who has contributed most to global CO2 emissions? Our World data (2019). Available:. Available from: https://ourworldindata.org/contributed-most-global-co2.

[4] IRENA (2019), "FUTURE OF WIND, Deployment, investment, technology, grid integration and socio-economic aspects (A Global Energy Transformation paper)," International Renewable Energy Agency, Abu Dhabi, 2019, Available: https://www.irena.org/-/media/Files/IRENA/Agency/Publication/2019/Oct/IRENA_Future_of_wind_2019.pdf.

[5] IEA, Renewables 2020 Data Explorer, IEA, Paris, 2020. Available from: https://www.iea.org/articles/renewables-2020-data-explorer.

[6] M. Kamran, et al., Solar Photovoltaic Grid Parity: A Review of Issues, Challenges, and Status of Different PV Markets, Int. J. Renew. Energy Res. 9 (1) (2019) 244–260.

[7] H. Bahar, "Solar PV," International Energy Agency (IEA), June 2020, Available: https://www.iea.org/reports/solar-pv.

[8] H. Bahar, "Onshore Wind," International Energy Agency (IEA), June 2020, Available: https://www.iea.org/reports/onshore-wind.

[9] H. Bahar, "Offshore Wind," International Energy Agency (IEA), June 2020, Available: https://www.iea.org/reports/offshore-wind.

Chapter 2

Thermodynamics for renewable energy systems

Muhammad Kamran
Department of Electrical Engineering and Technology, Riphah International University, Pakistan

2.1 Introduction

Thermodynamics is defined as the science of energy. In almost all renewable energy sources, thermodynamics is involved in one or the other way. The first law of thermodynamics governs the conservation of energy. In renewable energy sources, energy transforms from biomass to electrical, solar to electrical, solar to thermal, wind to electrical, biogas to electrical, biogas to thermal, chemical (hydrogen) to electrical, and the potential energy of water to electrical. In all these conversions, energy does transform its type but the total energy of the universe is conserved.

The second law of thermodynamics governs the entropy of the system saying that the universe tends to be more disordered. The refrigerators and the heat engines follow the second law of thermodynamics. The third law of thermodynamics states that the entropy of the system tends to approach zero as the temperature approaches zero kelvin.

In renewable energy system applications, thermodynamics is involved in solar thermal applications like solar tower, solar parabolic troughs, and solar dish. The Rankine cycle and the Brayton cycles govern these solar thermal applications. In hydrogen production and its utilization in fuel cells, enthalpy and entropy are involved to completely understand the working of the fuel cell.

2.2 Thermodynamic system

Any specific area that is under consideration with some thermodynamic properties of interest is called a system. Anything that is out of the system is called the surroundings and the limit between the system and the surroundings is called the boundary of the system. The boundary can be a real or an imaginary

Renewable Energy Conversion Systems. DOI: https://doi.org/10.1016/B978-0-12-823538-6.00004-X

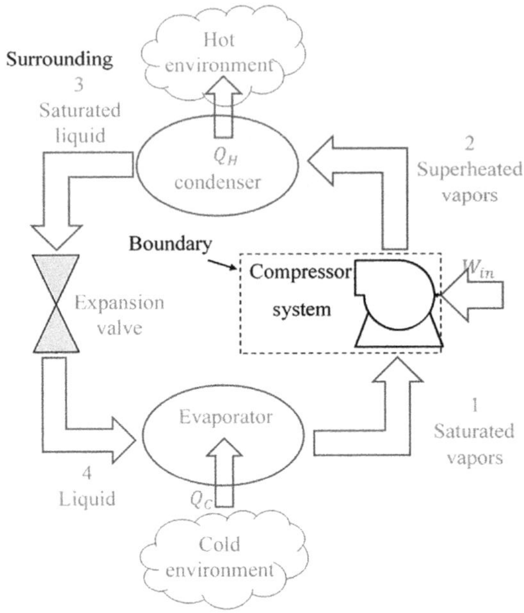

FIGURE 2.1 System, boundary, and surroundings.

surface separating the system and the surroundings. Consider a vapor–compression cycle of a refrigerator shown in Fig. 2.1. The compressor is under observation, where we are interested in studying the temperature, pressure, and efficiency of the compressor. The compressor is the system and everything except the compressor is the surroundings and the imaginary line that limits the system and the surroundings is the boundary of the system.

2.2.1 Open system

An open system is a type of thermodynamic system that can exchange energy as well as mass with its surroundings. Boiling water is an example of an open system in which the heat is going out of the vessel and the steam as a mass also goes out of the system to the surroundings. Fig. 2.2A shows the example of an open thermodynamic system.

2.2.2 Closed system

A closed system is a type of thermodynamic system that can exchange heat with its surroundings but mass cannot be exchanged. A pressure cooker is an example of a closed system. The energy/heat can be transferred to the surroundings, but the mass within the cooker cannot be transferred. Fig. 2.2B shows the example of a closed thermodynamic system.

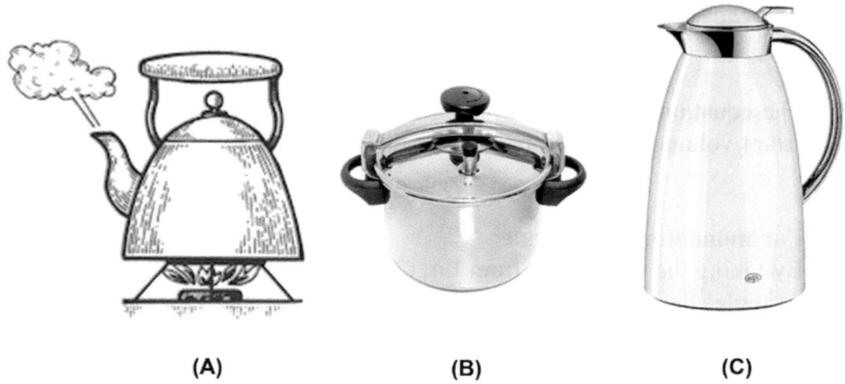

FIGURE 2.2 (A) Open system, (B) closed system, (C) isolated system.

2.2.3 Isolated system

An isolated system is a system that neither exchanges the heat with the environment nor can the mass can transferred. A thermos flask is the application of the isolated system. A thermos flask is used to keep things warm or cold, so heat can neither go in nor out of the flask. Fig. 2.2C shows the example of an isolated thermodynamic system [1].

2.3 Heat capacity

Heat capacity, also known as the specific heat, is defined as the amount of heat in joules required to raise the temperature of a 1 g substance by 1°C without changing the phase of the substance. It determines how much temperature has been changed by an absorbed amount of heat. Mathematically it is defined as the amount of heat divided by the temperature change, as shown in Eq. (2.1), where n is the number of moles of the substance.

$$C = n\frac{dQ}{dT} \tag{2.1}$$

Water has the highest heat capacity, which means that it requires a large amount of heat to raise its temperature which makes it feasible to use as heat storage in renewable energy systems. Water in a well-insulated container is used as a heat storage substance that takes the heat from solar collectors and is used for space heating, hot water necessities, and process heating.

2.3.1 Heat capacity at constant volume (C_V)

Specific heat or the heat capacity of a substance is not a constant quantity as it depends upon the temperature change, volume, and pressure. Therefore heat capacity is determined at constant pressure and constant volume. The equation for heat capacity at constant volume (C_V) can be written as Eq. (2.2).

$$Q = nC_V\Delta T \tag{2.2}$$

The equation of the first law of thermodynamics for an isochoric process (constant volume) has been reduced to Eq. (2.3).

$$Q = \Delta U \tag{2.3}$$

For monoatomic molecule

By taking the value of Q from Eq. (2.2), and the value of ΔU for monoatomic molecules from the equipartition principle of energy, we can find the heat capacity of a monoatomic molecule at constant volume, as given in Eq. (2.4) and Eq. (2.5).

$$nC_V\Delta T = \frac{3}{2}nR\Delta T \tag{2.4}$$

$$C_V = \frac{3}{2}R \tag{2.5}$$

where R is the general gas constant.

For diatomic molecule

Similarly, by taking the value of Q from Eq. (2.2), and the value of ΔU for diatomic molecules from the equipartition principle of energy, we can find the heat capacity of a diatomic molecule at constant volume, as given in Eq. (2.6) and Eq. (2.7).

$$nC_V\Delta T = \frac{5}{2}nR\Delta T \tag{2.6}$$

$$C_V = \frac{5}{2}R \tag{2.7}$$

2.3.2 Heat capacity at constant pressure (C_P)

Heat capacity at constant pressure is higher than heat capacity at constant volume because at constant pressure the heat added to the system do some pressure-volume work and increases the temperature of the system. In heat capacity at constant volume, all the heat added to the system is used to raise the temperature. Eq. (2.1) for heat capacity at constant pressure (C_P) can be written as Eq. (2.8).

$$Q = nC_P\Delta T \tag{2.8}$$

The equation of the first law of thermodynamics for an isobaric process (constant pressure) is the following Eq. (2.9).

$$Q = \Delta U + W \tag{2.9}$$

For monoatomic molecule

By taking the value of Q from Eq. (2.8), and the value of ΔU for monoatomic molecules from the equipartition principle of energy, and the W from the general gas equation we can find the heat capacity of a monoatomic molecule at constant volume, as given in Eq. (2.10) and Eq. (2.11).

$$nC_P\Delta T = \frac{3}{2}nR\Delta T + nR\Delta T \tag{2.10}$$

$$C_P = \frac{5}{2}R \tag{2.11}$$

For diatomic molecule

Similarly, By taking the value of Q from Eq. (2.8) and the value of ΔU for diatomic molecules from the equipartition principle of energy, we can find the heat capacity of a diatomic molecule at constant volume, as given in Eq. (2.12) and Eq. (2.13).

$$nC_P\Delta T = \frac{5}{2}nR\Delta T + nR\Delta T \tag{2.12}$$

$$C_P = \frac{7}{2}R \tag{2.13}$$

The ratio of the heat capacity at constant pressure to the heat capacity at constant volume is called the heat capacity ratio and is denoted by $\frac{C_P}{C_V} = \gamma$. The heat capacity ratio for the monoatomic gas is $\frac{5}{3}$ and for diatomic gas it is $\frac{7}{5}$.

2.3.3 Mayer's equation

A German physicist, Julius Robert Mayer presented that specific heat at constant pressure is greater than the specific heat at constant volume because the energy that is given to the system has not only increased the temperature but also has to perform some work. We take the value of Q from Eq. (2.8), ΔU from Eq. (2.2) and Eq. (2.3), W from the general gas equation, and put them in the equation of the first law of thermodynamics.

$$Q = \Delta U + W$$

$$nC_P\Delta T = nC_V\Delta T + nR\Delta T$$

$$C_P - C_V = R \tag{2.14}$$

Eq. (2.14) is called Mayer's equation. Mayer's equation can be verified for both monoatomic and diatomic gases.

2.4 Phase change and latent heat

To change the phase of a substance from solid to liquid to gas, we require energy. This energy is used to break the chemical bonding and overcome the bonding forces. If we measure the temperature during the phase change we come to know that despite the heat addition, the temperature of the substance does not change. This could be explained by the fact that during the phase change the heat added to the system is used to break the bonding of the molecules, which why the temperature does not change. Once the phase change is completed, the temperature starts increasing [2].

2.4.1 Latent heat of fusion

The energy that is absorbed by the substance when its phase changes from solid to liquid is called the latent heat of fusion. Equivalently, it is the heat that is liberated during a phase change from liquid to solid. One gram of ice at 0°C needs 334 J energy to melt to water. After taking 334 J energy, solid ice is converted to liquid water but its temperature remains 0°C. This heat is called the latent heat of fusion, which changes the phase of solid to liquid, and this heat is used in breaking the chemical bonds of the solid substance. The amount of heat required to change the phase of the substance for a given mass is determined by multiplying the latent heat of fusion by the mass, as shown in Eq. (2.15).

$$Q = mL_f \tag{2.15}$$

where m is mass of the substance, and L_f is the latent heat of fusion.

2.4.2 Latent heat of evaporation

The energy that is absorbed by the substance when its phase changes from liquid to gas is called the latent heat of evaporation. Equivalently, it is the heat that is liberated during a phase change from gas to liquid. One gram of water at 100°C needs 22.64 J of energy to evaporate to steam. After taking 22.64 J energy, liquid water is converted to steam but its temperature remains 100°C. This heat is called the latent heat of evaporation, which changes the phase of liquid to gas, and this heat is used in breaking the chemical bonds of the liquid substance. The amount of heat required to change the phase of the substance from liquid to gas for a given mass is determined by multiplying the latent heat of evaporation to the mass as shown in Eq. (2.16).

$$Q = mL_v \tag{2.16}$$

where m is the mass of the substance, and L_v is the latent heat of fusion.

From the discussion of latent heat of water, it can be seen that the latent heat of fusion is greater than the latent heat of evaporation. The reason behind this is the strong bonding forces in a solid where the molecules are tightly bound with each other and a large amount of heat is required to break these forces. On the other hand, in liquids, molecules are weakly bound with each other and a relatively lower amount of energy is needed to break the bonds.

Example 2.1

During the winter season, the water on the microhydro turbine froze at the temperature $-17°C$. The engineers suggested that the ice must be removed from the turbine by steaming it out at the temperature of 115°C otherwise the blades of the turbine would be rusted. The estimated mass of water is 5000 g.

Given data

To calculate the amount of heat that is required to melt the ice and then evaporate it by steaming at 115°C, we need the following data from the literature.

$$\text{Latent heat of fusion of water} = L_f = 80\ \text{cal/g}$$

$$\text{Latent heat of fusion of water} = L_v = 540\ \text{cal/g}$$

$$\text{Heat capacity of ice} = C_{P,ice} = 0.50\ \text{cal/g}\cdot{}^{\circ}\text{C}$$

$$\text{Heat capacity of water} = C_{P,water} = 1\ \text{cal/g}\cdot{}^{\circ}\text{C}$$

$$\text{Heat capacity of steam} = C_{P,steam} = 0.48\ \text{cal/g}\cdot{}^{\circ}\text{C}$$

Solution

When the heat Q_1 is added to the system, the ice starts melting and its temperature decreases until 0°C.

$$Q_1 = m{\cdot}C_{P,ice}{\cdot}\Delta T$$

$$Q_1 = 5000 \times 0.50 \times (0 + 17)$$

$$Q_1 = 42.5\ \text{kcal}$$

When Q_2 is added to the system at 0°C, the temperature remains at 0°C despite the addition of Q_2 to the system. This is the transition phase from solid to liquid and the heat added during this phase is called the latent heat of fusion.

$$Q_2 = m{\cdot}L_f$$

$$Q_2 = 5000 \times 80$$

$$Q_2 = 400\ \text{kcal}$$

When the heat Q_3 is added to the system, the temperature increases from 0°C to 100°C, and the temperature changes from 0°C to 100°C.

$$Q_3 = m \cdot C_{P,water} \cdot \Delta T$$

$$Q_3 = 5000 \times 1 \times (100 - 0)$$

$$Q_3 = 500\,\text{kcal}$$

When Q_4 is added to the system at 100°C, the temperature holds at 100°C despite the addition of Q_4 to the system. This is also the transition phase from liquid to steam and the heat added during this phase is called the latent heat of evaporation.

$$Q_4 = m \cdot L_v$$

$$Q_4 = 5000 \times 540$$

$$Q_4 = 2700\,\text{kcal}$$

In the final stage, Q_5 is added to the system and during this phase steam is formed and the temperature rises to 115°C (Fig. 2.3).

$$Q_5 = m \cdot C_{P,water} \cdot \Delta T$$

$$Q_5 = 5000 \times 0.48 \times (115 - 100)$$

$$Q_5 = 36\,\text{kcal}$$

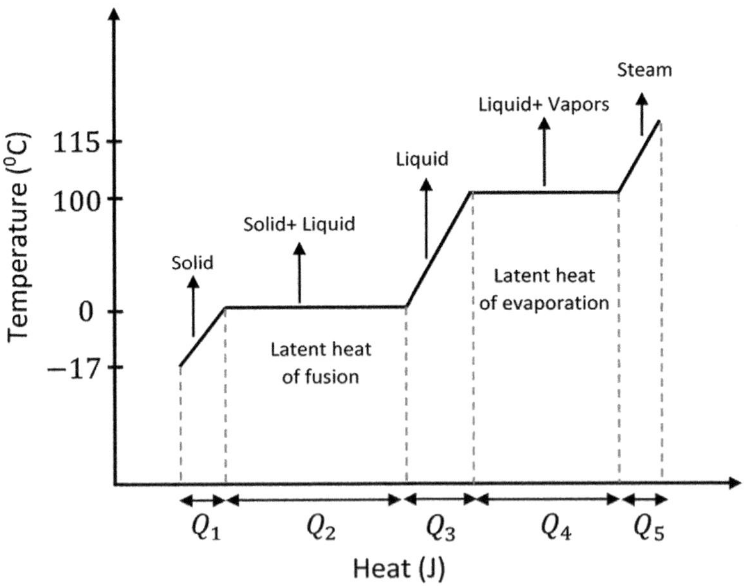

FIGURE 2.3 Phase change from ice to steam.

By determining the Q_1, Q_2, Q_3, Q_4, Q_5 and adding them all we can find the total heat added to the system.

$$Q_{total} = Q_1 + Q_2 + Q_3 + Q_4 + Q_5$$

$$Q_{total} = 3678.5\,\text{kcal}$$

2.5 Zeroth law of thermodynamics

Zeroth law of thermodynamics tells us whether the heat will flow between two bodies or not. The zeroth law of thermodynamics states that if a body A is in thermal equilibrium with another body B, and body A is also in thermal equilibrium with a body C, then this implies that the bodies B and C are also in equilibrium with each other. When two bodies of different temperatures are brought close to each other, the heat starts flowing from the high-temperature body to the lower-temperature body until both attain the same temperature and the heat flow process stops. At this point, the two bodies are said to be in thermal equilibrium. The process of the zeroth law of thermodynamics is shown in Fig. 2.4.

A very common example of the zeroth law of thermodynamics is the mercury thermometer. When the bulb of the thermometer is brought in contact with a hot body, the heat starts to flow from the body to the bulb, thus increasing the temperature of the mercury. The mercury expands in the column giving the measurement of the temperature. As soon as the temperature of the mercury and the body becomes the same, the heat transfer stops and both the hot body and the thermometer come into thermal equilibrium. Another example of the zeroth law of thermodynamics is the thermostat of the air conditioner of your room. When the room temperature becomes equal to the set temperature of the thermostat, they are in thermal equilibrium with each other and the compressor of the air conditioner is tripped OFF.

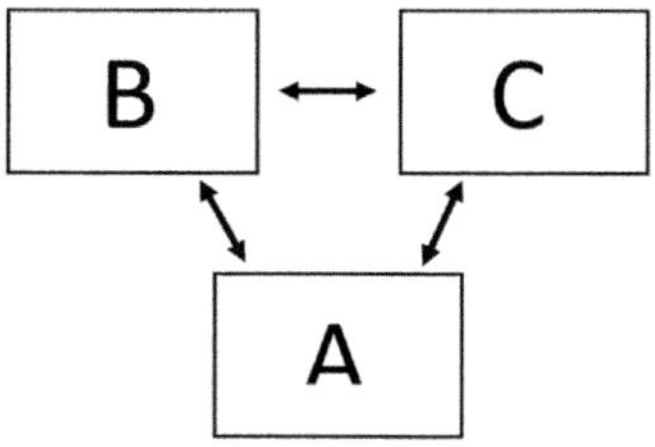

FIGURE 2.4 Zeroth law of thermodynamics.

2.6 The first law of thermodynamics

Generally, the first law of thermodynamics states that energy can neither be created nor destroyed, however it can be changed from one form to another. The total energy of the universe remains the same. For example, we cannot create kinetic energy to run the water turbine. We can convert the potential energy of water into kinetic energy that will be used to drive the turbine [3]. Thermodynamically the first law thermodynamics can be stated as when a specific amount of heat is added to a system, a part of that energy is used to do some useful work, while the rest is used in raising the internal energy of the system. Mathematically it can be written as

$$Q = \Delta U + W \tag{2.17}$$

From the above equation, we can define the change in internal energy as the energy added to the system minus the work done by the system. Mathematically we can write in Eq. (2.18),

$$\Delta U = Q - W \tag{2.18}$$

2.6.1 Isothermal process

A thermodynamic process that occurs at constant temperature is termed the isothermal process. As the heat slowly goes in or comes out of the system, it seems to be in equilibrium. Consider a system working at a temperature T_1, volume V_1, and pressure P_1 at time t_1. Keeping the temperature of the system constant, we add some heat Q to the system that increases the volume of the system to V_2 and decreases the pressure of the system to P_2 at time t_2. As the temperature of the system is kept constant ($\Delta T = 0$) so there is no change in its internal energy ($\Delta U = 0$). If we put $\Delta U = 0$ in the equation of the first law of thermodynamics we get $Q = W$ which describes that all the heat added to the system has been used to do some external work [4]. For this isothermal process, work done can be calculated as follows.

To calculate the work done by a system during a process is determined by Eq. (2.19).

$$W = P{\cdot}V \tag{2.19}$$

For an isothermal process work can be calculated by finding the area under the pressure–volume diagram shown in Fig. 2.5 and calculated in Eq. (2.20) as follows:

$$W = \int_{V_1}^{V_2} P{\cdot}dV \tag{2.20}$$

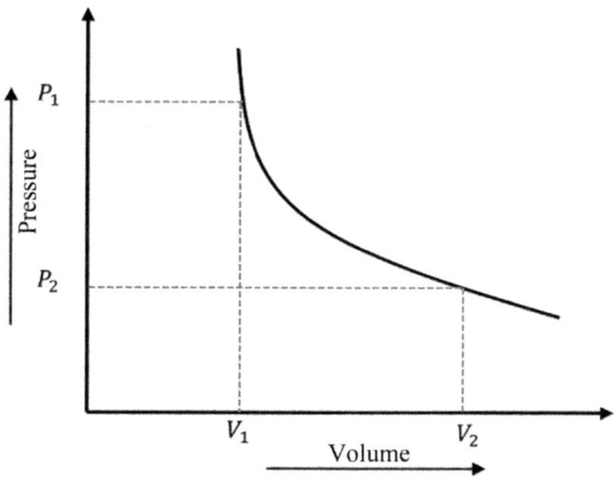

FIGURE 2.5 Work done in an isothermal process.

By general gas equation, we know the following relation $P = \frac{nRT}{V}$

$$W = \int_{V_1}^{V_2} \frac{nRT}{V} \cdot dV$$

$$W = nRT \int_{V_1}^{V_2} \frac{1}{V} \cdot dV$$

$$W = nRT \cdot \ln\left(\frac{V_2}{V_1}\right) \quad (2.21)$$

Eq. (2.21) is used for the calculation of the work done in an isothermal process.

Example 2.2

A mole of an ideal gas sealed in a cylinder is isothermally expanded to five times to its original volume 1 m^3 at a temperature of 290 K by slowly transferring heat into the gas.

1. What is the initial and final pressure of the gas?
2. How much heat is transferred into the gas?

Solution

$$n = 0.1 \text{ mole}$$

$$V_i = 1 \text{ m}^3$$

$$V_f = 5 \text{ m}^3$$

$$T = 290 \text{ K}$$

1. Initial and final pressures of the gas can be found by the general gas equation ($PV = nRT$)

$$P_i = \frac{nRT}{V_i}$$

$$P_i = \frac{(0.1\ \text{mol})\left(8.314\ \text{J/mol}\cdot\text{K}\right)(290\ \text{K})}{1\ \text{m}^3}$$

$$P_i = 241\ \text{Pa}$$

Similarly, the general gas equation is used to calculate the final pressure of the gas at 5m^3.

$$P_f = \frac{nRT}{V_f}$$

$$P_f = \frac{(0.1\ \text{mol})\left(8.314\ \text{J/mol}\cdot\text{K}\right)(290\ \text{K})}{5\ \text{m}^3}$$

$$P_f = 48.22\ \text{Pa}$$

2. As the ideal gas was expanded isothermally, no change in internal energy occurs, that is, $\Delta U = 0$, hence $Q = W$. We use Eq. (2.21) to calculate work done in the isothermal process that would be equal to the amount of heat transferred into the gas.

$$Q = W = nRT\cdot\ln\left(\frac{V_f}{V_i}\right)$$

$$Q = (0.1\ \text{mol})\left(8.314\ \text{J/mol}\cdot\text{K}\right)(290\ \text{K})\cdot\ln\left(\frac{5\ \text{m}^3}{1\ \text{m}^3}\right)$$

$$Q = 388\ \text{J}$$

2.6.2 Isobaric process

An isobaric process is a process occurring at constant pressure. The first law of thermodynamic equation for the isobaric process remains the same as the pressure remains constant and because of the volume change, the system does work. Work done can be calculated by the following equation. The PV diagram of the isobaric process is shown in Fig. 2.6.

$$W = P\cdot\Delta V = P(V_2 - V_1)$$

Eq. (2.17) of the first law of thermodynamics can be rewritten as Eq. (2.22).

$$Q = \Delta U + P(V_2 - V_1) \tag{2.22}$$

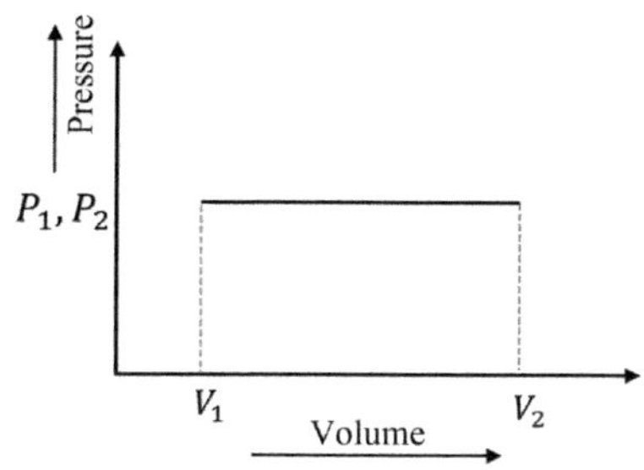

FIGURE 2.6 Work done in an isobaric process.

Example 2.3

One mole of an ideal diatomic gas, initially at 30°C and a constant pressure 2 bar is allowed to expand from 5 to 10 m^3 by giving heat at a temperature of 50°C. Find out the change in the internal energy of the gas.

Solution

$$Q = \Delta U + P(V_2 - V_1)$$

$$nC_P\Delta T = \Delta U + P(V_2 - V_1)$$

$$\Delta U = nC_P\Delta T - P(V_2 - V_1)$$

$$\Delta U = (2)\frac{(7)(8.314\ \text{J/mol}\cdot\text{K})}{2} - (2\text{bar})(323 - 303\ \text{K})$$

$$\Delta U = 18.2\ \text{J}$$

2.6.3 Isochoric process

An isochoric process is a process that happens at constant volume. Constant volume means there is zero work done by the system and Eq. (2.17) can be written as Eq. (2.23). It means all the heat energy provided to the system is used to raise the internal energy of the system. The PV diagram of the isochoric process is shown in Fig. 2.7.

$$Q = \Delta U \tag{2.23}$$

Example 2.4

One mole of an ideal diatomic gas, initially at 30°C and 2 bar, is compressed isothermally to a point such that when it is heated isochorically to 100°C its final pressure is 8 bar.

1. Calculate W, Q, and ΔU for the isothermal process.
2. Calculate W, Q, and ΔU for the isochoric process.

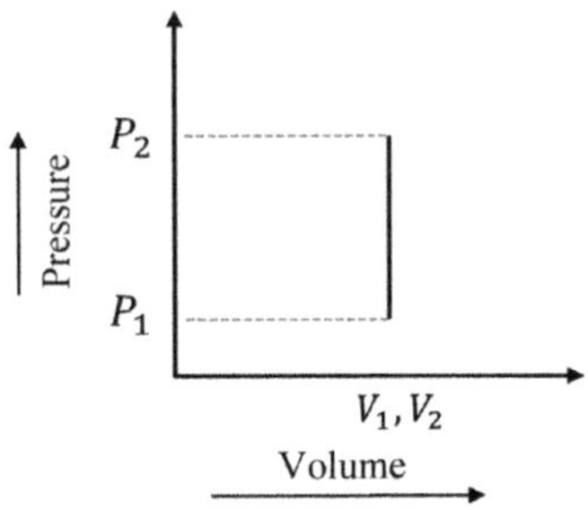

FIGURE 2.7 Work done in the isochoric process.

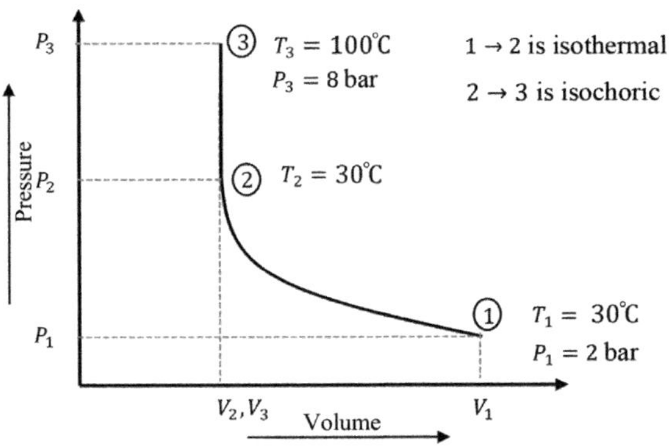

FIGURE 2.8 Isothermal-isochoric process for example 2.4.

Solution

The whole process consists of an isothermal process (1—2) and an isochoric process (2—3). We draw the graph of the given processes and label the required parameters on it and find out the missing parameters, like V_1, V_2, V_3, and P_2 (Fig. 2.8).

Pressure and volume of the gas at given three stages of the process can be found by the general gas equation ($PV = nRT$)

For V_1

$$P_1V_1 = nRT_1$$

$$V_1 = \frac{nRT_1}{P_1}$$

$$V_1 = \frac{(1)(R)(303)}{1}$$

$$V_1 = 303R$$

For V_2 and V_3

$$P_3V_3 = nRT_3$$
$$V_3 = \frac{nRT_3}{P_3}$$
$$V_3 = \frac{(1)(R)(303)}{8}$$
$$V_3 = 37.9R = V_2$$

For P_2

$$P_2V_2 = nRT_2$$
$$P_2 = \frac{nRT_2}{V_2}$$
$$P_2 = \frac{nRT_2}{V_2}$$
$$P_2 = \frac{(1)(R)(303)}{37.9R} = 7.99 \text{ bar}$$

1. Work done for the isothermal process ($1 \Rightarrow 2$) is calculated by using Eq. (2.21).

$$W = nRT\cdot\ln\left(\frac{V_2}{V_1}\right)$$
$$W = (1)(R)(303)\cdot\ln\left(\frac{37.9R}{303R}\right)$$
$$W = -630R \text{ J}$$

As the process is isothermal, so no change in internal energy takes place, that is, $Q = W$.

$$Q = -630R \text{ J}$$

2. Work done in the isochoric process ($2 \Rightarrow 3$) is zero as no change in volume takes place and the amount of heat transfer is the same as the change in internal energy which is calculated by using Eq. (2.2). We take $C_p = 7R/2$ and $C_v = 5R/2$ as the gas is diatomic.

$$Q = \Delta U = nC_V\Delta T$$
$$Q = \Delta U = (1)\left(\frac{5R}{2}\right)(373 - 303)$$
$$Q = \Delta U = 175R \text{ J}$$

2.6.4 Adiabatic process

A process in which a system does not exchange heat with its surroundings is termed an adiabatic process. During the adiabatic expansion, the system does work at the expense of its internal energy. As there is no exchange of heat, so $Q = 0$. Eq. (2.17) of the first law of thermodynamics can be written as Eq. (2.24).

$$0 = \Delta U + W$$

$$W = -\Delta U \tag{2.24}$$

Eq. (2.24) indicates that without any exchange of heat or mass, the system did work by using its own internal energy. Eq. (2.24) can be rewritten as Eq. (2.26) by taking the value of W from Eq. (2.19) and internal energy from Eq. (2.2) and Eq. (2.3), and we can develop a relationship for the adiabatic work.

$$P{\cdot}\Delta V = -nC_V\Delta T$$

$$dT = -\frac{P{\cdot}dV}{nC_V} \tag{2.25}$$

By differentiating the general gas equation with respect to temperature and putting the value of dT from Eq. (2.25), we get the following relation.

$$P{\cdot}dV + V{\cdot}dP = -\frac{R}{C_V}(P{\cdot}dV)$$

By putting the value of the general gas constant R from Mayer's equation, and dividing the whole relation with PV we get the following relation.

$$\frac{dV}{V} + \frac{dP}{P} = -\left(\frac{C_P}{C_V} - 1\right)\frac{dV}{V}$$

Where $\frac{C_P}{C_V} = \gamma$ is called the heat capacity ratio and its value for monoatomic and diatomic gas molecules is determined.

$$PV^{\gamma} = Constant \tag{2.26}$$

Eq. (2.26) is a well-established fact.; during an adiabatic expansion, the product of pressure and V^{γ} remains constant.

Example 2.5

A certain diatomic ideal gas is expanded adiabatically; initially, 2 mol of gas is at a pressure of $101{,}300/\text{m}^2$, the volume of 5 m^3 and temperature of 25°C. The final volume increases to 10 m^3. Calculate the final pressure of the gas.

Given data

$$n = 2\text{ mol}$$
$$P_1 = 101,300\frac{N}{\text{m}^2}$$
$$V_1 = 5\text{ m}^3$$
$$V_2 = 10\text{ m}^3$$
$$T_1 = 25°\text{C}$$

Solution

As the gas is expanded adiabatically, Eq. (2.26) reveals that the product PV^γ remains constant at any time during the reaction, that is,

$$P_1V_1^\gamma = P_2V_2^\gamma$$
$$P_2 = P_1\left(\frac{V_1}{V_2}\right)^\gamma$$

where $\gamma = \frac{C_P}{C_V}$ is the capacity ratio and is calculated for diatomic gas using Eq. (2.7) and Eq. (2.13)

$$\gamma = \frac{7R/2}{5R/2} = \frac{7}{5}$$
$$P_2 = 101,300\frac{N}{\text{m}^2}\left(\frac{5\text{ m}^3}{10\text{ m}^3}\right)^{7/5}$$
$$P_2 = 38385.5\ \frac{N}{\text{m}^2}$$

Example 2.6

A gas cylinder holds 2 mol of O_2 at 125°C and a pressure of 2.0 atm. The gas expands adiabatically until the pressure is 6.0 atm. What are the final volume and temperature?

Given data

$$n = 2\text{ mol}$$
$$P_1 = 3\text{ atm}$$
$$P_2 = 6\text{ atm}$$
$$T_1 = 125°\text{C}$$
$$V_1 = ?$$
$$V_2 = ?$$
$$T_2 = ?$$

Solution

We find the initial volume of the diatomic gas by using the general gas equation.

$$P_1V_1 = nRT_1$$

$$V_1 = \frac{nRT_1}{P_1}$$

$$V_1 = \frac{(2)(8.31)(398)}{3\,(101,300)}$$

$$V_1 = 0.022\ \text{m}^3$$

As the gas is expanded adiabatically, Eq. (2.26) reveals that the product PV^γ remains constant at any time during the reaction, that is,

$$P_1V_1^\gamma = P_2V_2^\gamma$$

$$3(0.022)^{7/5} = 6(V_2)^{7/5}$$

$$V_2 = 0.0134\ \text{m}^3$$

The final temperature of the gas can be calculated by the fact that the general gas equation remains constant at any time of the process, that is,

$$\frac{P_1V_1}{T_1} = \frac{P_2V_2}{T_2}$$

$$T_2 = \frac{P_2V_2T_1}{P_1V_1}$$

$$T_2 = \frac{(6\ \text{atm})\left(0.0134\ \text{m}^3\right)(398\ \text{K})}{(3\ \text{atm})(0.022\ \text{m}^3)}$$

$$T_2 = 485\ \text{K}$$

2.7 The second law of thermodynamics

2.7.1 Kelvin–Planck statement

The Kelvin–Planck statement of the second law of thermodynamics, also known as the heat engine statement, states that it is impossible to devise a heat engine that takes heat from the hot reservoir (Q_H) and converts all the energy into useful external work without losing heat to the cold reservoir (Q_C). For the above process to happen, a part of the input energy must be wasted in the environment. In other words, the thermal efficiency of a heat engine cannot be 100%. The Kelvin–Planck statement of the second law of thermodynamics can also be explained by Fig. 2.9.

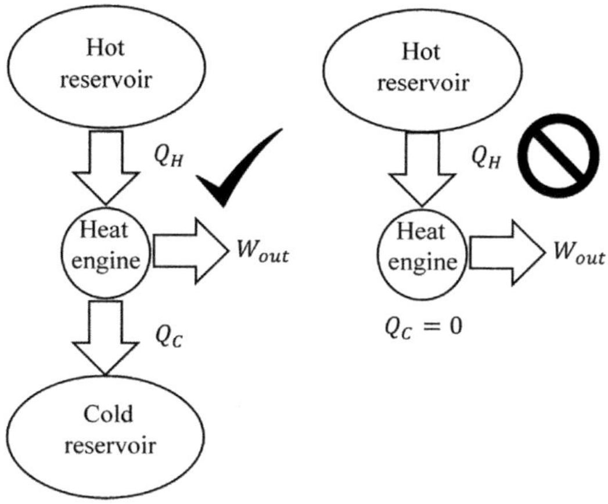

FIGURE 2.9 The Kelvin–Planck statement of the second law of thermodynamics.

If we consider the statement of Fig. 2.9 to be true and put the value $Q_C = 0$ in Eq. (2.27), we will have a heat engine that would be 100% efficient, which is not possible according to the Kelvin–Planck statement of the second law of thermodynamics. Eq. (2.27) is used to determine the efficiency of a heat engine.

$$\eta = \frac{Output}{input} 100\%$$

$$\eta = \frac{W_{out}}{Q_H} 100\%$$

$$\eta = \frac{Q_H - Q_C}{Q_H} 100\%$$

$$\eta = \left(1 - \frac{Q_C}{Q_H}\right) 100\% \tag{2.27}$$

$$\eta = 100\%$$

So the 100% efficiency negates the possibility of having a system converting all the heat of the hot reservoir into useful work.

Example 2.7

In a biomass power plant, 100 MW of heat is transferred from a furnace to the gas turbine. 30 MW is converted into the mechanical energy of the coupled generator. Find:

1. The amount of heat being rejected to the cold reservoir.
2. The thermal efficiency of the gas turbine.

Given data

$$Q_H = 100\,\text{MW}$$
$$W_{out} = 30\,\text{MW}$$

Solution

1. By the definition of the Kelvin–Planck statement of the second law of thermodynamics, we know that

$$W_{out} = Q_H - Q_C$$

$$Q_C = 100\,\text{MW} - 30\,\text{MW}$$

$$Q_C = 70\,\text{MW}$$

2. The efficiency of the turbine is calculated using Eq. (2.27)

$$\eta = \left(1 - \frac{Q_C}{Q_H}\right)100\%$$

$$\eta = \left(1 - \frac{70}{100}\right)100\%$$

$$\eta = 30\%$$

The turbine can convert only 30% of the heat into useful work, whereas the remaining 70% is wasted in the atmosphere.

2.7.2 Clausius statement

The Clausius statement of the second law of thermodynamics, also known as the heat engine or the refrigerator statement, says that it is impossible to construct a system that takes heat Q_C from the cold reservoir at temperature T_C and moves it to the hot reservoir at temperature T_H without external work. It is a well-established concept of physics that heat moves spontaneously from higher potential to low and the reverse is not possible without some external work done. The Clausius statement of the second law of thermodynamics can also be explained by Fig. 2.10.

As the second law of thermodynamics discusses the refrigeration, air conditioning, and heat pump processes, so we discuss here the complete vapor–compression cycle that works in these processes. Fig. 2.11 shows the complete vapor–compression refrigeration cycle. The purpose of the vapor–compression cycle is to take the heat from the cold environment of the refrigerator and throw it out into the warm environment. The

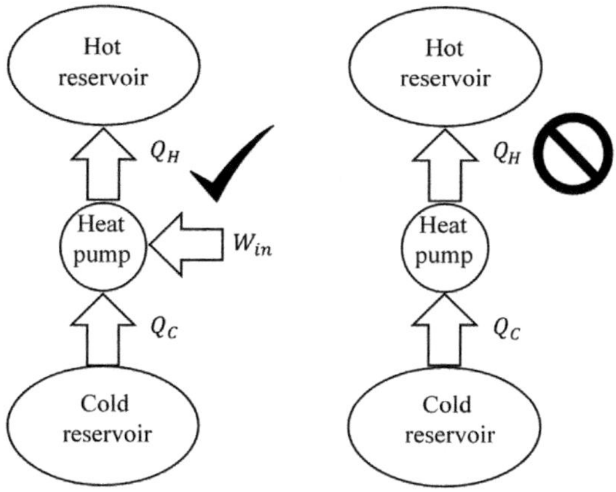

FIGURE 2.10 The Clausius statement of the second law of thermodynamics.

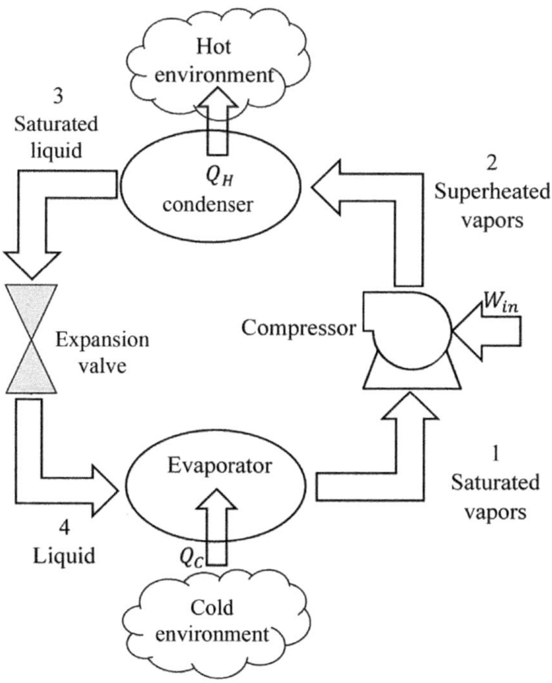

FIGURE 2.11 A refrigeration cycle.

vapor–compression cycle consists of four parts performing the following four thermodynamic processes:

1. Compressor; adiabatic process
2. Condenser; isobaric process

3. Expansion valve; adiabatic process
4. Evaporator; isobaric process

A refrigerant is used that takes the heat from the hot reservoir and throws it out by changing its phase throughout the above four processes. At the first stage from phase 1 to phase 2 in Fig. 2.11, the refrigerant in saturated vapor form is compressed in a compressor by external work. After that from phase 2 to phase 3, the superheated vapors are condensed to saturated liquid in a condenser at constant pressure, and heat is rejected to the hot environment. From phase 3 to phase 4, the saturated liquid is expanded in an expansion valve that decreases the pressure. From phase 4 to phase 1 the liquid refrigerant is evaporated by taking the heat from the cold environment. After evaporation, the saturated liquid is converted into vapors and these vapors still contain energy in the form of heat. Concisely, the liquid refrigerant takes heat from the cold environment in an evaporator and it is converted to vapors. These vapors are condensed in a condenser where the vapors are converted to liquid and the heat is rejected to the hot environment.

The efficiency of the refrigerator is not measured by the output to input ratio. It is measured by the actual objective of the refrigerator which is the amount of heat Q_C extracted from the cold reservoir. Coefficient of performance (*COP*) is introduced to measure the performance of the refrigerator which is defined by Eq. (2.28).

$$COP = \frac{Desired\ output}{Required\ input} \tag{2.28}$$

$$COP = \frac{Q_C}{W_{in}} = \frac{h_1 - h_4}{h_2 - h_1} \tag{2.29}$$

From Fig. 2.11 we can see that $W_{in} = Q_C - Q_H$. Eq. (2.29) can be rewritten as Eq. (2.30)

$$COP = \frac{Q_C}{Q_C - Q_H} \tag{2.30}$$

$$COP = \frac{1}{Q_H/Q_C - 1} \tag{2.31}$$

From Eq. (2.31) we can grasp the concept that unlike the efficiency, *COP* can be greater than 1. It means that the amount of heat extracted from the cold reservoir can be greater than the work done by the compressor.

Example 2.8

A refrigerator requires 3 kW to maintain the internal temperature at 10°C by removing heat at a rate of 220 kJ/min. Find:

1. The rate of heat rejection from inside the refrigerator to the environment.
2. The *COP* of the refrigerator.

Given data

$$Q_C = 220\,\text{kJ/min}$$
$$W_{in} = 3\,\text{kW}$$

Solution

1. By the definition of the Kelvin–Planck statement of the second law of thermodynamics, we know that

$$W_{out} = Q_H - Q_C$$

$$3\text{kW}\left(\frac{60\,\text{kJ/min}}{1\text{kW}}\right) = Q_H - 220\,\text{kJ/min}$$

$$Q_H = 400\,\text{kJ/min}$$

2. The *COP* of the refrigerator is calculated using Eq. (2.31)

$$COP = \frac{1}{Q_H/Q_C - 1}$$

$$COP = \frac{1}{\left(400\dfrac{\text{kJ}}{\text{min}} \Big/ 220\dfrac{\text{kJ}}{\text{min}}\right) - 1}$$

$$COP = 1.22$$

Example 2.9

A refrigerator working on the ideal vapor–compression cycle uses 134-a as a working fluid. The pressure at the input of the compressor and the output of the expansion valve is 700 kPa, whereas the pressure at the output of the compressor and input of the expansion valve is 160 kPa. The mass flow rate of the refrigerant is 0.1 kg/second. Find:

1. Power required by the compressor.
2. The heat taken from the refrigerator.
3. Heat rejected to the environment.
4. *COP* of the refrigerator.

Given data

$$P_1 = P_4 = 160\,\text{kPa}$$
$$P_2 = P_3 = 700\,\text{kPa}$$
$$\dot{m} = 0.1\,\text{kg/s}$$

Solution

From Fig. 2.11 and the given data, we can deduct the following information from the pressure table of the refrigerant 134-a.

$$\boldsymbol{P_1 = 160\,kPa}$$

$$\Rightarrow h_1 = h_g = 241.14\,\text{kJ/kg}$$

$$and\;\; s_1 = s_g = 0.94202\,\text{kJ/kg}\cdot\text{K}$$

$$\boldsymbol{P_2 = 700\,kPa}$$

$$\Rightarrow s_2 = s_1 = 0.94202\,\text{kJ/kg}\cdot\text{K}$$

$$and\;\; h_2 = 273.54\,\text{kJ/kg}$$

$$\boldsymbol{P_3 = 700\,kPa}$$

$$\Rightarrow h_3 = h_f = 88.82\,\text{kJ/kg}$$

$$h_4 \cong h_3 = 88.82\,\text{kJ/kg}$$

1. The power that is required to compress the refrigerant in a compressor is determined by the following equation

$$W_{in} = \dot{m}(h_2 - h_1)$$

$$W_{in} = 0.1\ \text{kg/s}\left(273.54\frac{\text{kJ}}{\text{kg}} - 241.14\ \text{kJ/kg}\right)$$

$$W_{in} = 3.24\ \text{kW}$$

2. The heat taken from the refrigerator is determined as follows

$$Q_c = \dot{m}(h_1 - h_4)$$

$$Q_c = 0.1\ \text{kg/s}(241.14\ \text{kJ/kg} - 88.82\ \text{kJ/kg})$$

$$Q_c = 15.23\ \text{kW}$$

3. Heat rejected to the environment is determined by the following equation

$$Q_H = \dot{m}(h_2 - h_3)$$

$$Q_H = 0.1\ \text{kg/s}(273.54\ \text{kJ/kg} - 88.82\ \text{kJ/kg})$$

$$Q_H = 18.47\ \text{kW}$$

4. *COP* of the refrigerator is determined by the following relation

$$COP = \frac{Q_c}{W_{in}} = \frac{15.23\ \text{kW}}{3.24\ \text{kW}}$$

$$COP = 4.7$$

2.8 Third law of thermodynamics

The third law of thermodynamics states that the entropy of the system tends to approach zero as the temperature approaches zero kelvin.

2.9 Thermodynamic cycles

2.9.1 Solar thermal Brayton cycle (GAS)

Gas cycles operate in an open system in which air from the atmosphere is taken and compressed in a pump that raises the pressure and temperature of the air. The compressed air is passed through the receiver of the solar tower where the heat energy of concentrated radiation of the sun is transferred to the compressed air at constant pressure. The compressed hot gases are then passed through the turbine coupled to the generator. The hot gases are expanded to rotate the turbine and then exhaust out at atmospheric pressure. Fig. 2.12 shows the thermodynamic solar thermal Brayton cycle as described earlier. As the exhausted gases are not recirculated this cycle is termed the open cycle [5].

Fig. 2.13 shows the modification model in the open cycle and is termed as the close cycle. The direct thermal process is replaced by an isobaric heat addition process and the exhaust process is replaced by an isobaric heat rejection process. This cycle is called the closed-loop solar thermal Brayton cycle consisting of

1–2 Isentropic compression in the pump by external work.
2–3 Isobaric heat addition in evaporator from the solar receiver.
3–4 Isentropic expansion in turbine doing output work.
4–1 Isobaric heat rejection in the heat exchanger.

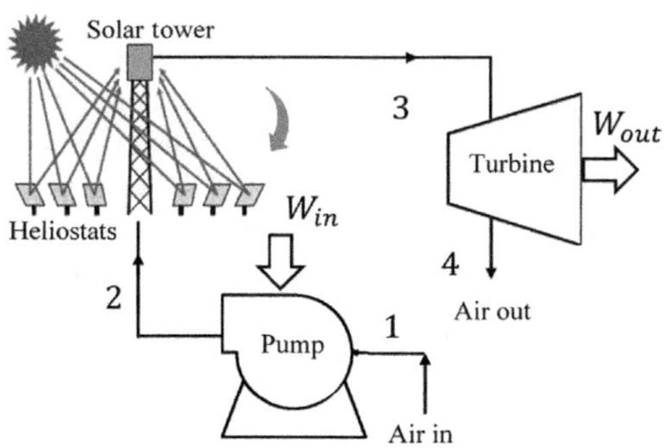

FIGURE 2.12 Solar thermal open Brayton cycle.

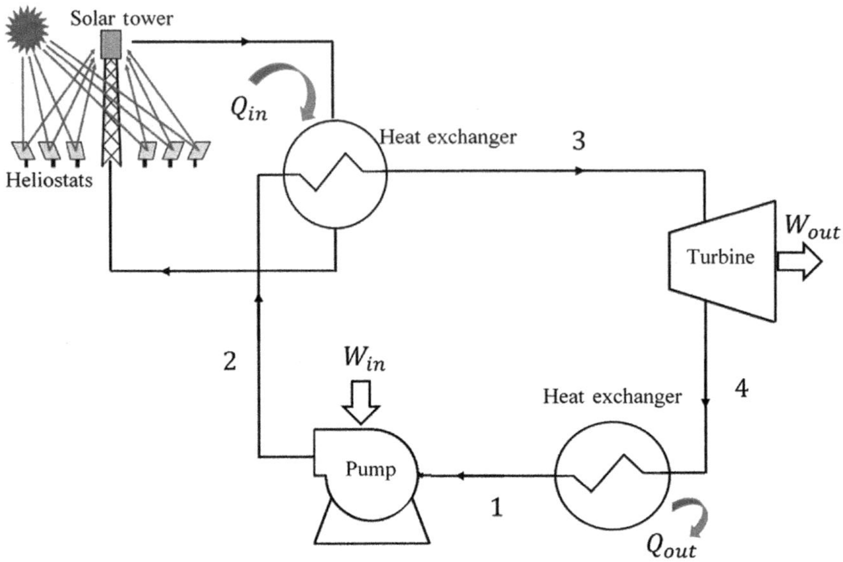

FIGURE 2.13 Solar thermal close Brayton cycle.

2.9.2 Solar thermal organic Rankine cycle (STEAM)

For steam power plants, the Rankine cycle is considered to be ideal. It consists of the following four thermodynamic cycles. Organic Rankine cycle (ORC) is used for low-temperature applications in renewable energy. At present, its share in renewable energy is 1% solar thermal, 20% waste heat recovery, 31% geothermal, and 48% biomass. We discuss here the Rankine cycle for solar thermal power plants. As the ORC is used for low-temperature availability, it contains a low boiling point working organic fluid [6]. Some of the working liquids are R-601 (n-pentane), R-123, R-134a, benzene (C6H6), R-245fa, R-717 (ammonia), and R-601a (isopentane). In the discussion, we use a parabolic trough with a solar collector, where the parabolic trough concentrates the solar radiation onto the solar collector. The solar collector contains a working fluid that takes the heat and transfers its heat to the water in the evaporator. The water in the evaporator takes the heat and generates the steam for the steam power plant. The ORC process with the heat transfer fluid mechanism for the solar thermal power plant is shown in Fig. 2.14. For renewable energy power plants to be operated for steam power stations, we can replace the solar parabolic trough and the solar dish with biomass and geothermal [7].

1−2 Isentropic compression in the pump by external work.

2−3 Isobaric heat addition in evaporator from the solar collector.

3−4 Isentropic expansion in turbine doing output work.

4−1 Isobaric heat rejection in condenser to cooling tower.

At state 1 water is compressed isentropically in a pump that raises the pressure of the water to the operating point of the evaporator. In the evaporator

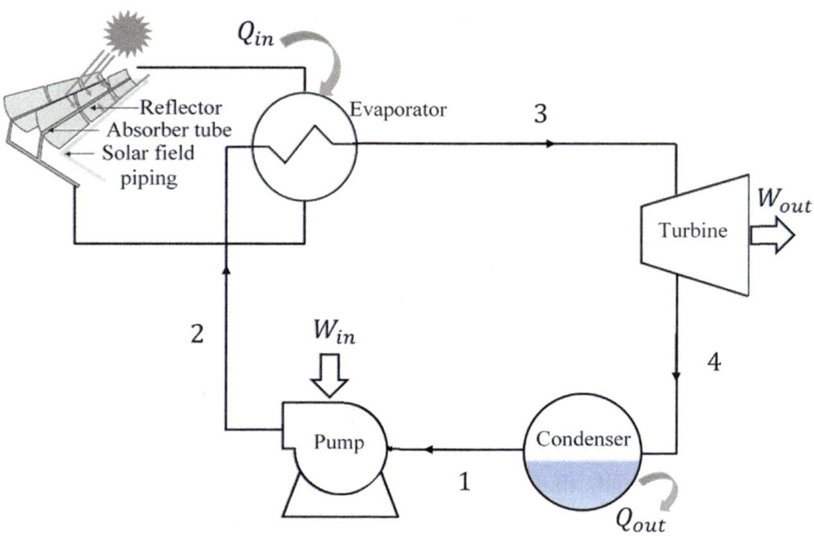

FIGURE 2.14 Solar thermal Rankine cycle.

compressed water takes the heat from the working fluid of the parabolic trough and is converted to superheated steam. The evaporator is like a heat exchanger that takes the heat from the solar collector. At state 3 the superheated steam is expanded on the turbine that rotates and runs an electrical generator. During expansion at the turbine, the temperature and the pressure of the steam drops and the steam is sent to the condenser. In the condenser, the liquid–vapor mixture is condensed and the saturated liquid is sent to the compressor. The condenser behaves like a heat exchanger that rejects the heat to the cooling tower. The process continues in the same fashion.

For the energy analysis of the ORC, we can determine that there is no heat transfer involved in the compressor and the turbine where the isentropic compression and expansion occur respectively. The pump requires the input work done to operate while the turbine does external work. On the other hand, the evaporator takes the heat from the solar collector and the condenser rejects the heat to the environment. No work done is involved in the evaporator and the condenser. The enthalpy change during each process in each device can be stated as follows:

$$W_{in} = h_2 - h_1 \tag{2.32}$$

$$Q_{in} = h_3 - h_2 \tag{2.33}$$

$$W_{out} = -(h_4 - h_3) \tag{2.34}$$

$$Q_{out} = -(h_1 - h_4) \tag{2.35}$$

Eq. (2.32) is used to determine the work done on the compressor in terms of the change of enthalpy across the pump. Eq. (2.33) determines the heat transfer from the solar thermal to the evaporator in terms of enthalpy change across the evaporator. Eq. (2.34) gives the work done by the turbine, whereas Eq. (2.35) shows the amount of heat rejected by the condenser. Eq. (2.36) gives the thermal efficiency of the organic Rankine cycle.

$$\eta = \frac{output\ energy}{input\ energy}$$

$$\eta = \frac{W_{out} - W_{in}}{Q_{in}}$$

$$\eta = \frac{Q_{in} - Q_{out}}{Q_{in}}$$

$$\eta = \left(1 - \frac{Q_{out}}{Q_{in}}\right) 100\% \tag{2.36}$$

Eq. (2.36) reveals that the efficiency of the ORC depends upon the heat addition in the evaporator and the heat rejection in the condenser.

2.9.3 Solar combined power cycle

Solar combined power cycles are gaining popularity over the conventional combined cycle power plants. These are replacing the gas-based combustion

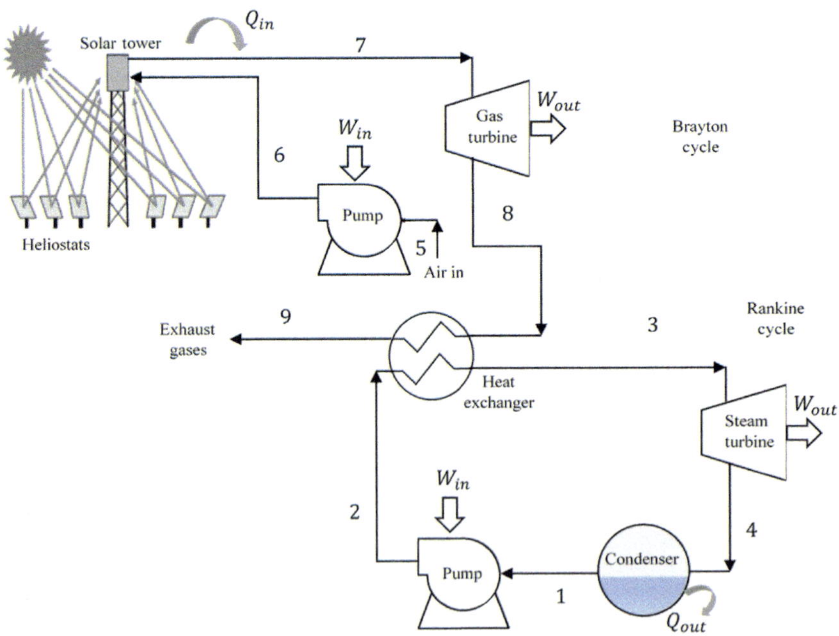

FIGURE 2.15 Solar combined power cycle.

with the solar thermal and the reduction in potential emissions. In the solar combined power cycle, the combustion chamber is replaced by a concentrated solar power system consisting of heliostats and the solar tower. Heliostats reflect the solar radiation to the solar receiver on the solar tower. In the upper Brayton cycle, the compressed air is passed through the solar tower that converts the air into hot gases. These hot gases are expanded over the gas turbine and the flue gases are passed through the heat exchanger where the heat is used to generate steam in the bottom Rankine cycle [8]. Fig. 2.15 shows the solar thermal combined power cycle.

Problems

Problems 1–10 contains four answer options A, B, C, and D. Choose the correct answer.

1. Thermodynamics can be defined as the
- **A.** Science of Chemistry
- **B.** Science of Physics
- **C.** Science of energy
- **D.** Science of Mathematics

2. ---------------- separates the thermodynamic system from the surroundings.
- **A.** Mass
- **B.** Energy
- **C.** Boundary
- **D.** Heat

3. An open system exchanges ------------------ with its surroundings.
- **A.** Mass
- **B.** Energy
- **C.** Mass and energy
- **D.** Heat

4. A closed system exchanges ------------------ with its surroundings.
- **A.** Mass
- **B.** Energy
- **C.** Mass and energy
- **D.** Heat

5. Heat capacity of a diatomic gas at constant volume is
- **A.** $\frac{5}{2}R$
- **B.** $\frac{3}{2}R$
- **C.** $\frac{7}{2}R$
- **D.** R

6. Mayer's equation can be defined as
- **A.** $C_P - C_V = R$
- **B.** $Q = \Delta U + W$
- **C.** $\Delta U = Q - W$
- **D.** All of above

7. Which is exactly true for an isothermal process?
   A. $\Delta U = 0$
   B. $Q = W$
   C. $\Delta U = 0$ and $Q = W$
   D. All of above
8. Which is exactly true for an isochoric process?
   A. $\Delta U = 0$
   B. $Q = W$
   C. $Q = \Delta U$
   D. $\Delta U = 0$ and $Q = W$
9. Which statement defines the second law of thermodynamics?
   A. Kelvin–Planck statement
   B. Clausius statement
   C. Gibbs statement
   D. Mayer's statement
   E. A & B
   F. C & D
10. Which of the following thermodynamic cycle is for steam?
    A. Rankine cycle
    B. Otto cycle
    C. Brayton cycle
    D. Carnot cycle
11. Water cannot be used as a fuel to run the car engine. Give a reason in the light of second law of thermodynamics.
12. Explain the latent heat of fusion and latent heat of evaporation when ice is converted into steam. Also write down the phase changes involved in this process.
13. What are the cycles that are involved in the combined cycle? State the flow process of the combined cycle.
14. Briefly explain the difference between a refrigerator, a heat engine, and a heat pump.

References

[1] G.G. Láng, Basic Thermodynamics of Electrified Interfaces, K.B.T.-E. of, I.C. Wandelt (Eds.), Elsevier, Oxford, 2018, pp. 1–14.

[2] I. Dincer, Y. Bicer, Chapter 2—Fundamentals of energy systems, in: I. Dincer, Y. B. T.-I. E. S. for, M. Bicer (Eds.), Integrated Energy Systems for Multigeneration, Elsevier, 2020, pp. 33–83.

[3] R.F. Sekerka, 2—First law of thermodynamics, in: R.F.B.T.-T.P. Sekerka (Ed.), Thermal Physics, Elsevier, Amsterdam, 2015, pp. 15–30.

[4] J.B. Ott, J. Boerio-Goates, Chapter 11—Summary of thermodynamic relationships, in: J.B. Ott, J.B.T.-C.T.A.A. Boerio-Goates (Eds.), Chemical Thermodynamics: Advanced Applications, Academic Press, London, 2000, pp. 1–42.

[5] S. Basu, A.K. Debnath, Chapter I—Introduction, in: S. Basu, A.K.B.T.-P.P.I., C.H. Debnath (Eds.), Power Plant Instrumentation and Control Handbook, Academic Press, Boston, 2015, pp. 1–37.

[6] Z. Dai, 17—Thorium molten salt reactor nuclear energy system (TMSR), in: T. J. B. T.-M. S. R., T.E. Dolan (Eds.), Molten Salt Reactors and Thorium Energy, Woodhead Publishing, 2017, pp. 531–540.

[7] M. Orosz, R. Dickes, 16—Solar thermal powered organic Rankine cycles, in: E. Macchi, M.B.T.-O.R.C. (O.R.C.) P.S. Astolfi (Eds.), Organic Rankine Cycle (ORC) Power Systems, Woodhead Publishing, 2017, pp. 569–612.

[8] B.J. Alqahtani, D. Patiño-Echeverri, Integrated solar combined cycle power plants: paving the way for thermal solar, Appl. Energy 169 (2016) 927–936. Available from: https://doi.org/10.1016/j.apenergy.2016.02.083.

Chapter 3

Power electronics for renewable energy systems

Muhammad Kamran
Department of Electrical Engineering and Technology, Riphah International University, Pakistan

3.1 Introduction

In the late 1880s, Thomas Edison was in favor of the direct current (DC) transmission system while Nicola Tesla was in favor of alternating current (AC) transmission. The war between AC and DC was won by Tesla because of the presence of transformers that could easily step up and step down the voltage levels of AC for transmission and distribution purposes. The other reason was the absence of DC circuit breakers. Now the AC transmission and distribution is the convention.

Now with the advancement in power electronics, we have power electronics equipment for conditioning of the voltages. Apart from the conventional semiconductor devices, we have now advanced power electronics switches like silicon controlled rectifier (SCR), silicon controlled switch (SCS), insulated gate bipolar transistors, gate turn-off thyristor (GTO), Diode for Alternating Current (DIAC), and Triode for Alternating Current (TRIAC). To step up and step down the DC voltage levels, DC–DC boost converters, Cuk converters, DC–DC buck converters, Single Ended Primary Inductor Converter (SEPIC) converters, flyback converters, and buck-boost converters are available. DC–DC converters are integrated with a control circuit of Maximum Power Point Tracker (MPPT) to let the solar photovoltaic (PV) operate at its maximum power point. To convert an AC signal into a DC signal we have AC–DC half-wave and full-wave rectifiers. For short circuit protection, we do have DC breakers available. DC–AC inverters are available in the market and are the backbone of the hybrid renewable energy system and distributed generation.

As the renewable energy sources are intermittent, power electronics is involved in solar energy, wind energy, fuel cell, geothermal, bioenergy, and hydropower to regulate the output electrical energy.

Renewable Energy Conversion Systems. DOI: https://doi.org/10.1016/B978-0-12-823538-6.00001-4

3.2 Solid-state devices

3.2.1 Silicon controlled rectifier (Thyristor)

The thyristor known as silicon controlled rectifier (SCR) is a four-layer p-n-p-n, three-junction J_1, J_2, J_3, and three-terminal, named as anode (A), cathode (K), and Gate (G), device. The thyristor has been widely used in power converters and hybrid renewable energy applications. It is used as a fast switching device in converters. SCR normally behaves like a forward biased diode if its anode is made positive with respect to the cathode and the control signal at the gate terminal is present. If the gate signal is absent, SCR is in forward blocking mode. Fig. 3.1 shows the layers, junctions, and the terminals of the SCR.

The two-transistor model shown in Fig. 3.2 presents the comprehensive understanding of the working of SCR. Two transistors are connected in such a way that both share their p-terminals and n-terminals. The base of Q_1, which is an n-terminal, is connected to the n-terminal collector of Q_2. Similarly the base of Q_2, which is a p-terminal, is connected to the p-terminal collector of Q_1. One left p-terminal of Q_1 is considered as ther anode and one left n-terminal of Q_2 is considered as the cathode. The shared p-terminal is considered as the control terminal, known as a gate.

In the first case we do not apply any control signal to the gate, hence there is no ignition to turn ON the Q_2. A positive voltage is applied at the anode with respect to the cathode that forward biases the emitter-base junction of Q_1 and Q_2 and reverse biases the collector-base junctions of Q_1 and Q_2. In the present state, though the two transistors are ON, there is no conduction through the SCR. Only a small leakage current flows through the reverse biased collector-base junction of Q_2, which is connected to the base

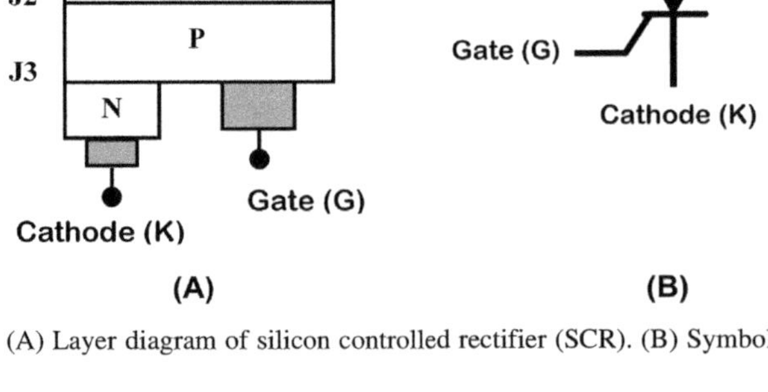

FIGURE 3.1 (A) Layer diagram of silicon controlled rectifier (SCR). (B) Symbol of SCR.

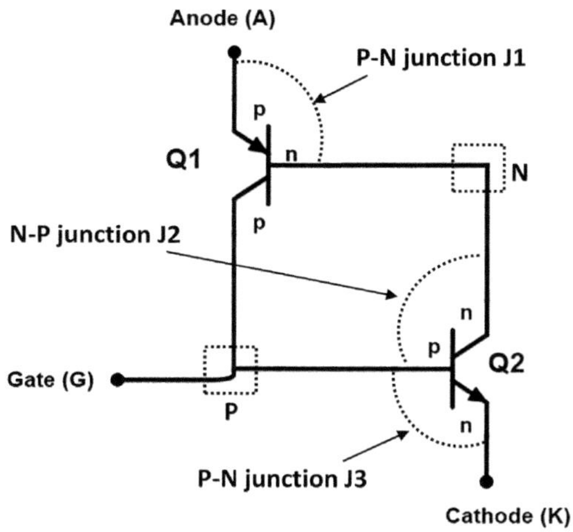

FIGURE 3.2 Two-transistor model of silicon controlled rectifier.

of Q_1. Likewise, a small leakage current flows through the reverse biased collector-base junction of Q_1 which is connected to the base of Q_2. These small leakage currents cannot start conduction through the two transistors. The total current through the SCR is the emitter current of Q_2, which is the sum of I_{B2} and I_{C2}. In the layer diagram shown in Fig. 3.1A, Junction J_1 and J_3 are forward biased and J_2 is still reverse biased so there is no conduction from anode to cathode, there is only a leakage current [1].

Now we increase the positive voltage at the anode that will increase the leakage current in the collector of Q_1, which in the present case is also the base current of Q_2. According to the relation $I_c = \beta I_B$, the collector current of Q_2 is increased, which is the base current of Q_1, which in return increases the collector current of Q_1. If we make the anode voltage large enough, that increases the collector current of Q_1 turning the Q_2 fully ON. The increased collector current of Q_2 will turn the Q_1 fully ON and the process will continue. The collector current flowing through one transistor Q_1 ensures full conduction through the other transistor Q_2 and the collector current of Q_2 ensures the conductivity through Q_1. This process is termed as the regenerative switching process. The voltage necessary to start the regenerative switching process is called the forward breakover voltage V_{BR}. As the V_{BR} is achieved, the voltage across the SCR is reduced and the current starts conducting as shown in Fig. 3.3. The device remains in conduction mode until its current is higher than the holding current. If the current falls below the holding current, the device returns to its forward blocking mode.

In the second case, the SCR is triggered by applying a short pulse at the gate terminal. As soon as the gate signal is applied, the reverse breakover voltage is reduced and the conduction starts abruptly. Once the device is latched, the gate current is no longer required and the device remains in

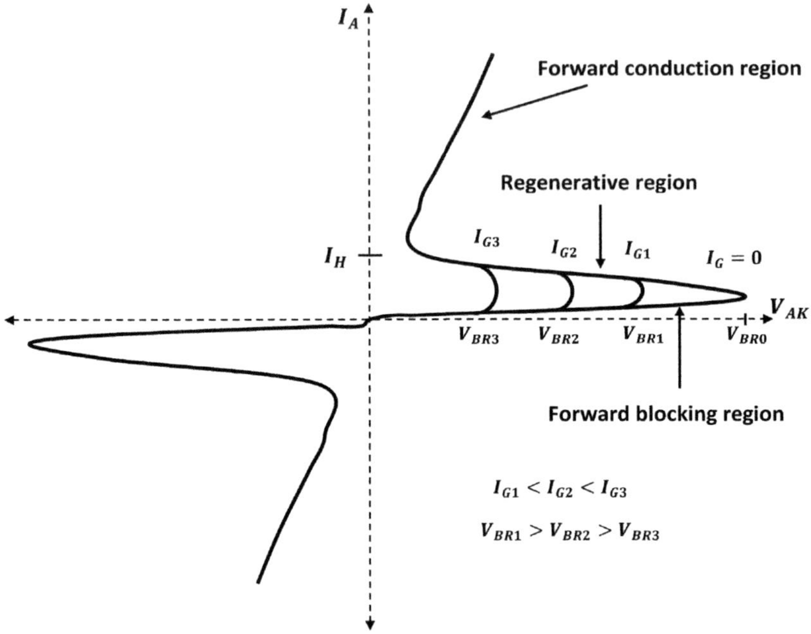

FIGURE 3.3 V–I characteristics of silicon controlled rectifier.

conduction mode. The value of V_{BR} depends upon the value of gate current. With the increase in gate current, the thyristor goes into conduction earlier and the voltage (V_{BR}) reduces across the thyristor, as shown in Fig. 3.3. The thyristor cannot be turned off by removing the gate signal, as it cannot bring the anode current below the holding current. To turn the thyristor OFF, the anode voltage is reduced such that the anode current falls below the holding current. Holding current is defined as the miimum value of the anode current below which the SCR goes into OFF mode or the forward blocking mode. The anode current is brought below the holding current either by natural commutation or the forced commutation. To again turn ON the SCR, we need to increase the anode current above the latching current. Latching current is the maximum value of the anode current through the forward biased diode that is required to bring the SCR into the forward conduction mode. Once the latch current is achieved, SCR is latched and the removal of gate signal does not turn OFF the device.

3.2.2 Gate turn-off thyristor

A gate turn-off thyristor (GTO) is a type of silicon controlled rectifier having three terminals—anode, cathode, and gate—with an extra feature of turning OFF by the gate terminal. The gate of the GTO is heavily doped with a

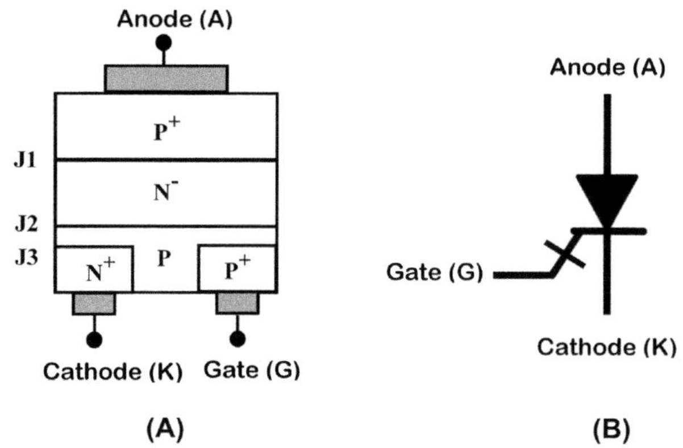

FIGURE 3.4 (A) Layer diagram of gate turn-off thyristor (GTO). (B) Symbol of GTO.

P-material. A GTO can be turned ON by a positive pulse at the gate and it can also be turned OFF by a negative pulse at the gate [2]. The negative pulse at the gate brings down the anode to cathode voltage below the V_{AK} that in turn reduces the current below the holding current turning it OFF. The structural diagram and the symbol of GTO are shown in Fig. 3.4A,B respectively.

3.2.3 Silicon controlled switch

The structure of silicon controlled switch (SCS) is like an SCR with an additional gate terminal. Like an SCR it can be turned ON by the primary gate terminal and like a conventional switch it can be turned OFF by the secondary gate terminal. The primary gate terminal is the common point of the p-collector of Q_1 and p-base of the Q_2 transistor and is termed the cathode gate. Similarly the secondary gate terminal is the common point of the n-base of Q_1 and n-collector of the Q_2 transistor and is termed the anode gate. Fig. 3.5A shows the layer diagram of the SCS where the anode gate is at the n-layer and the cathode gate is at the p-layer. Fig. 3.5B shows the symbol of the SCS.

The SCS can be turned ON by a positive current at the cathode gate that starts the regenerative process by driving Q_2 ON. On the same pattern SCS can also be turned on by applying a negative current at the anode gate that drives the Q_1 transistor and starts the regenerative process. SCS can be turned off by turning off either of the two transistors. Q_1 can be turned off by applying a positive current at the anode gate. Similarly Q_2 can be turned off by a negative current at the cathode gate. The ON and OFF process is shown in two-transistor models in Fig. 3.6.

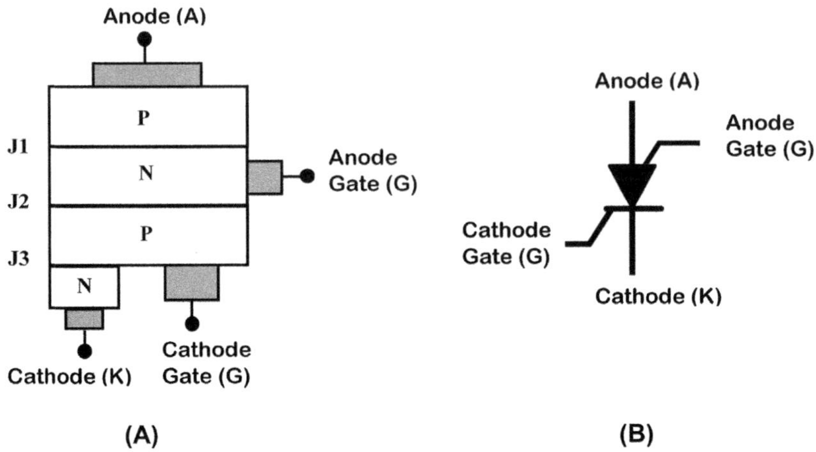

FIGURE 3.5 (A) Layer diagram of silicon controlled switch (SCS). (B) Symbol of SCS.

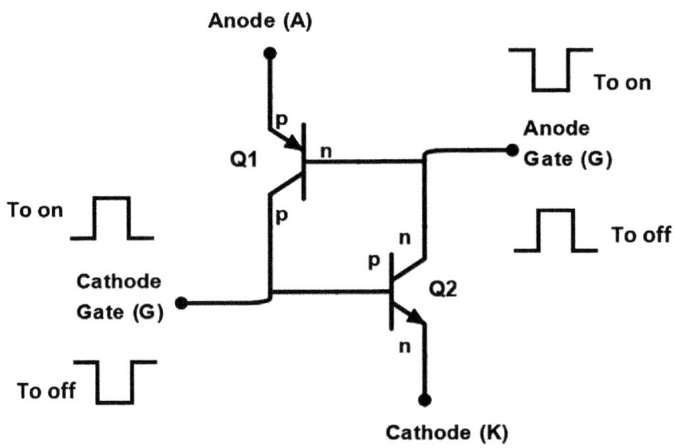

FIGURE 3.6 Two-transistor model of silicon controlled switch.

3.2.4 DIAC

A diode for alternating current (DIAC) consists of four layers, four junctions, and two terminals. From the layer diagram shown in Fig. 3.7A, it can be seen that a DIAC is a gateless device consisting of the stack of a gateless SCR and a gateless inverted SCR. The left side is the P-N-P-N structure and the right side is a N-P-N-P structure. There is no gate and the two terminals are labeled as A_1 and A_2. Fig. 3.7B shows the symbol of the DIAC.

Fig. 3.8 shows the transistor equivalent model of the DIAC. Q_1 and Q_2 represent gateless SCR and Q_3 and Q_4 represent the gateless inverted SCR. The p-emitters of Q_1 and n-emitter of Q_3 are connected to each other and

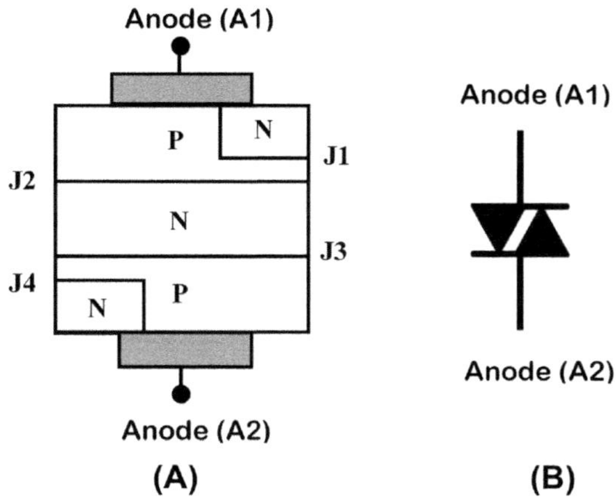

FIGURE 3.7 (A) Layer diagram of DIAC. (B) Symbol of DIAC.

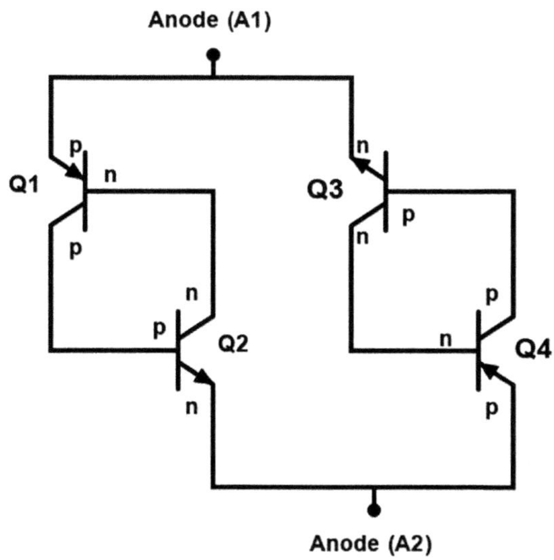

FIGURE 3.8 Equivalent transistor model of a DIAC.

this is termed as the anode 1 (A_1), whereas the n-emitters of Q_2 and p-emitter of Q_4 are connected to each other and this is termed as the anode 2 (A_2). DIAC is a bidirectional device conducting from either A_1 to A_2 or from A_2 to A_1 depending on the polarity of the anode voltage. If A_1 is made more positive than the A_2, Q_1 and Q_2 are ON, and Q_3 and Q_4 are OFF. Conduction starts from A_1 to A_2 through the transistors Q_1 and Q_2. To the contrary, if A_2 is made more positive than the A_1, Q_1 and Q_2 are OFF,

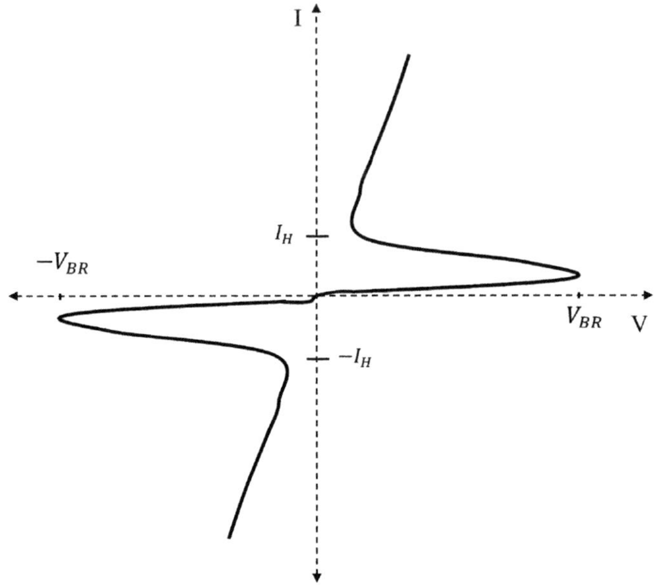

FIGURE 3.9 V–I characteristics of a DIAC.

whereas Q_3 and Q_4 are ON. Conduction starts from A_2 to A_1 through the transistors Q_4 and Q_3.

The V–I characteristic curve of the DIAC is shown in Fig. 3.9. The curve in the first quadrant indicates the conduction from A_1 to A_2 through Q_1 and Q_2. It is similar to the SCR curve without the gate signal. The curve in the third quadrant shows the reverse current flow from A_2 to A_1 through Q_4 and Q_3. The conduction starts when the voltage reaches the breakover voltage in the opposite polarity. The DIAC is turned OFF by bringing the current below the I_H if it is working in the first quadrant or below the $-I_H$ if it is in the third quadrant.

3.2.5 TRIAC

The basic structure of the triode for alternating current (TRIAC) is equivalent to the DIAC with an additional gate terminal. Fig. 3.10A shows the layer structure of the TRIAC, whereas Fig. 3.10B is the symbol of the TRIAC. The transistor equivalent model of the TRIAC is shown in Fig. 3.11. The gate terminal is at the combination of base terminals of Q_2 and Q_3, both of which are N-P-N transistors. The direction of the conduction between A_1 and A_2 depends upon the polarity of the gate signal with respect to A_1 and A_2. For conduction, two conditions need to be satisfied: polarity of the anode and the polarity of the gate terminal. If we need to drift the current from A_1 to A_2 through Q_1 and Q_2, the polarity of A_1 must be positive with respect to

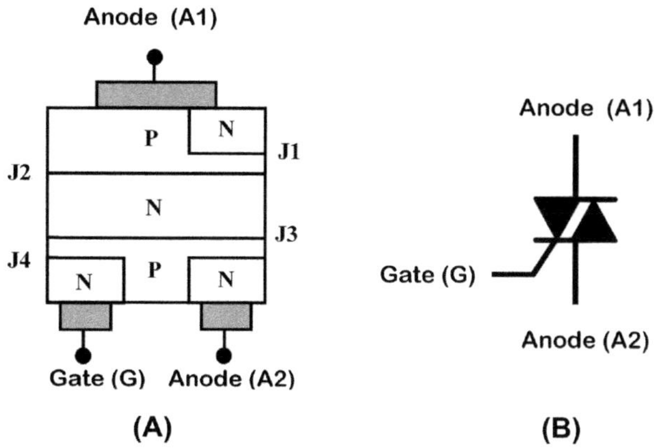

FIGURE 3.10 (A) Layer diagram of a TRIAC. (B) Symbol of a TRIAC.

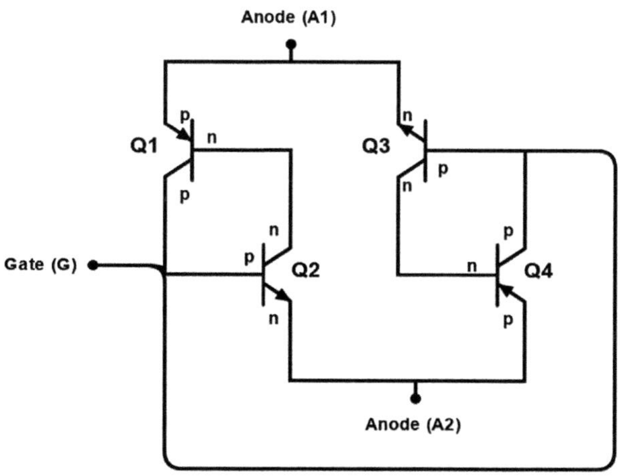

FIGURE 3.11 Equivalent transistor model of a TRIAC.

A_2, and the gate terminal must be positive with respect to A_2. To start the conduction from A_2 to A_1 through Q_3 and Q_4, A_2 must be made positive with respect to A_1 and the gate terminal must be positive with respect to A_1.

Fig. 3.12 shows the characteristic curves of the TRIAC for different values of gate currents in forward conduction mode in the first quadrant and reverse conduction mode in the third quadrant. As we increase the gate current, V_{BR} is achieved earlier and the conduction starts rapidly. Reverse conduction mode characteristic curves also indicate the same procedure for the reverse conduction of the current. In both quadrants to turn OFF the TRIAC, the current through the TRIAC must be reduced below the holding current.

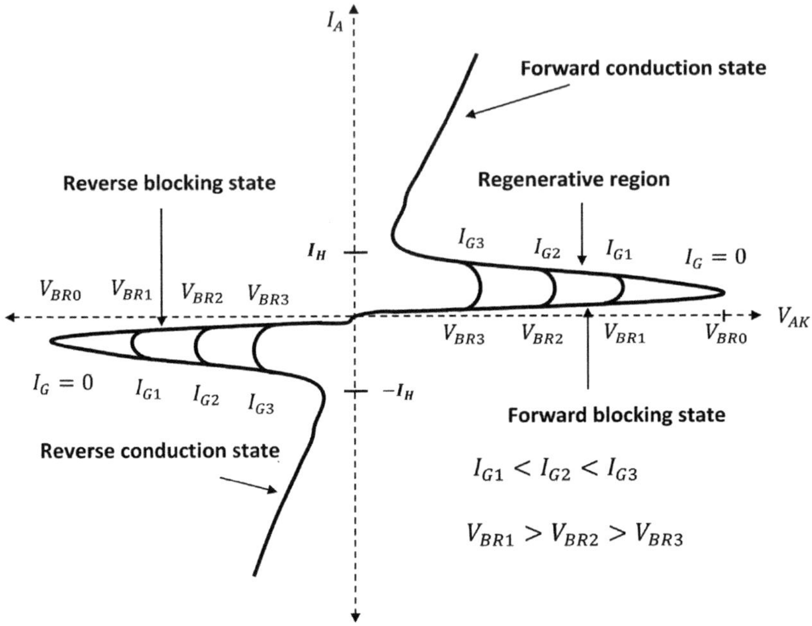

FIGURE 3.12 V–I characteristics of a TRIAC.

3.3 Rectifiers (AC–DC converters)

3.3.1 Half-wave uncontrolled rectifier with resistive load

Fig. 3.13 shows the half-wave uncontrolled rectifier with a resistive load. As the diode is a gateless or a baseless device having no control, all the input voltage during forward biasing appears across the output and all the input voltage during reverse biasing appears at the diode itself [3].

The input signal is AC that is fed to the output through the diode is shown in Eq. (3.1). During the first positive half cycle of the input AC signal, the diode is forward biased as the positive terminal is connected to the anode of the diode. The diode conducts for the input voltage for an ideal diode but in the actual case, the barrier potential is deducted from the input voltage. From Fig. 3.14 it can be seen that for the period 0–π, the voltage is dropped at the output resistor and no voltage is dropped across the diode. Similarly, for the period π–2π, the negative cycle of the input signal is there that reverse biases the diode and all the voltage is dropped across the diode and there is zero voltage at the output as shown in Fig. 3.14. The same process continues for the next cycle. Positive pulse voltage across the load and negative pulse voltage across the diode make the complete input cycle as illustrated in Fig. 3.14. As the load is resistive, the output current is in phase to the voltage.

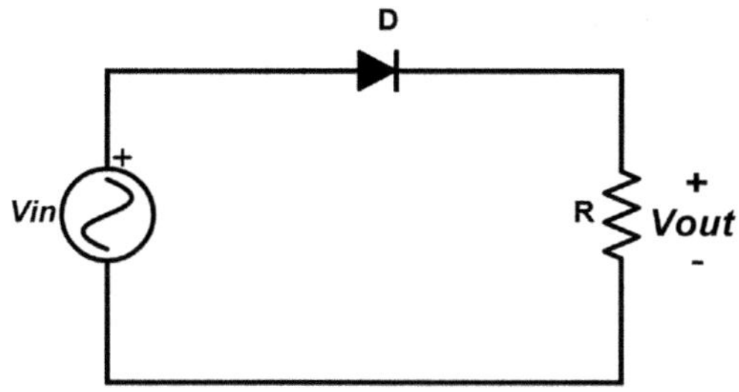

FIGURE 3.13 Half-wave uncontrolled rectifier with a resistive load.

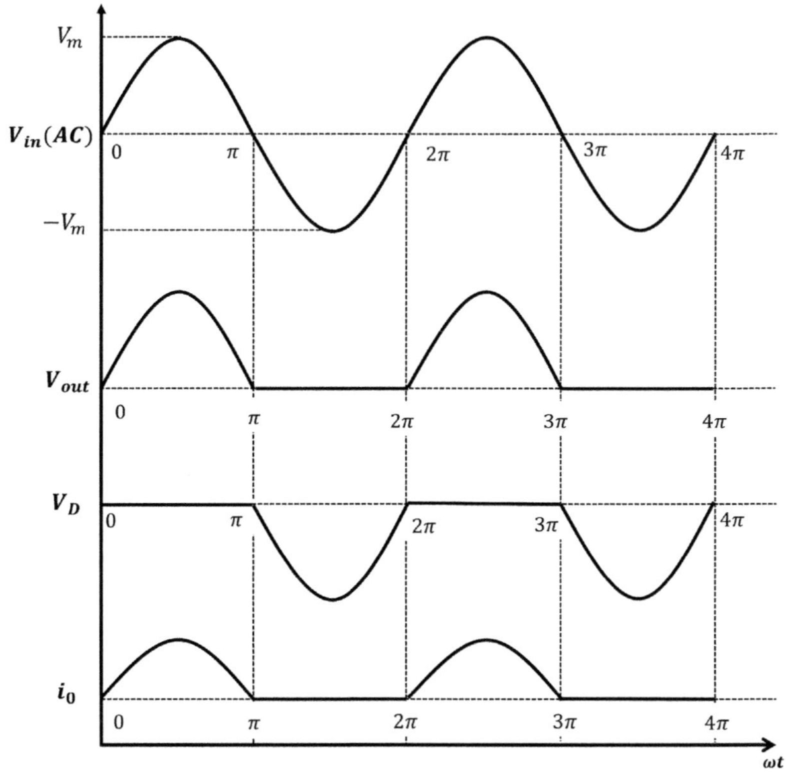

FIGURE 3.14 Output waveforms of a half-wave uncontrolled rectifier with a resistive load.

$$V_{in} = V_m \sin(\omega t) \tag{3.1}$$

The voltage at the output that is measured by a DC voltmeter is the average voltage which is mathematically calculated by dividing the area under one complete cycle by the time period.

$$V_{avg.} = \frac{\text{Area under the curve}}{T}$$

Area under the curve can be determined by integrating the complete cycle. We integrate only the first half cycle $(0-\pi)$, as in the second half cycle $(\pi-2\pi)$ we have no signal at the output. Eq. (3.2) is used to determine the average output voltage of the half-wave uncontrolled rectifier with a resistive load.

$$V_{O\ (avg)} = \frac{1}{2\pi}\int_0^{\pi} V_m \sin(\omega t)\ .\ d(\omega t)$$

$$V_{O\ (avg)} = \frac{V_m}{\pi} \tag{3.2}$$

The average value of the current is determined by Eq. (3.3).

$$I_{O\ (avg)} = \frac{V_{O\ (avg)}}{R} \tag{3.3}$$

The average power of the rectifier is determined by the multiplication of voltage and current given in Eqs. (3.2) and (3.3). Average output power of the half-wave uncontrolled rectifier with resistive load is determined by Eq. (3.4).

$$P_{O\ (avg)} = V_{O\ (avg)} \times I_{O\ (avg)}$$

$$P_{O\ (avg)} = \left(\frac{V_m}{\pi}\right)^2 \cdot \frac{1}{R} \tag{3.4}$$

Example 3.1

A half-wave rectifier is to be used to build a DC power supply for the battery charging system for the desktop computer UPS. The input AC voltage to the rectifier is $V_{in} = 80 \sin(\omega t)$, and the internal resistance of rectifier is 25 Ω. If the load resistance is 1 kΩ, calculate

(a) AC input power
(b) Average voltage at the output
(c) Average current at the output
(d) Average power at the output
(e) Efficiency of the rectifier

Solution
Given data

$$V_{in} = 80 \sin(\omega t)$$

By comparing the above equation of input voltage with Eq. (3.1), we can say that $V_m = 80\text{V}$

$$R_L = 25\Omega$$

(a) The input power to the rectifier will be calculated by considering the internal resistance of the forward biased diode in the following equation.

$$P_{in} = (I_{rms})^2 \times (r_f + R)$$

$$P_{in} = \left(\frac{I_m}{2}\right)^2 \times (r_f + R)$$

$$P_{in} = \left(\frac{I_m}{2}\right)^2 \times (r_f + R)$$

$$P_{in} = \left(\frac{V_m}{2(r_f+R)}\right)^2 \times (r_f + R)$$

$$P_{in} = \frac{V_m{}^2}{4(r_f + R)}$$

$$P_{in} = \frac{80^2}{4(25\Omega + 1k\Omega)}$$

$$P_{in} = 1.56W$$

(b) Average output voltage of the half-wave rectifier is determined by Eq. (3.2).

$$V_{O\ (avg)} = \frac{V_m}{\pi} \Rightarrow \frac{80}{\pi} = 25.47\ \text{V}$$

(c) We use Eq. (3.3) to determine the average output current.

$$I_{O\ (avg)} = \frac{V_{O\ (avg)}}{r_f + R} \Rightarrow \frac{25.47\ \text{V}}{25\ \Omega + 1\ k\Omega} = 24.8\ \text{mA}$$

(d) Average power at the output is determined by Eq. (3.4).

$$P_{O\ (avg)} = V_{O\ (avg)} \times I_{O\ (avg)}$$

$$P_{O\ (avg)} = 25.47\ \text{V} \times 24.8\ \text{mA}$$

$$P_{O\ (avg)} = 631\ \text{mW}$$

(e) Efficiency of the rectifier is determined by the ratio of the output DC power calculated in part (d) to the input AC power calculated in part (a).

$$\eta = \frac{P_{O\ (avg)}}{P_{in}} \Rightarrow \frac{631\ \text{mW}}{1.56\ \text{W}}$$

$$\eta = 40.4\%$$

3.3.2 Half-wave uncontrolled rectifier with inductive load

Fig. 3.15 shows the circuit diagram of a half-wave rectifier with an inductive load. As the diode is used as a rectifying element, it is called an uncontrolled rectifier. During the positive half cycle $(0-\pi)$, the diode is forward biased and all the input voltage appears across the load and zero voltage is across the diode, as shown in Fig. 3.16. From π onward, the input voltage starts increasing in the reverse direction making the diode reverse biased, but the current through the inductor starts decaying and voltage across it builds up with opposite polarity until reaching the angle ϕ. From ϕ to 2π, the diode is reverse biased and all the input voltage is dropped across the diode and zero voltage across the load. After that the cycle repeats itself. Because of the inductive load, the output current lags behind the output voltage.

By calculating the area under the output voltage for a period $(0\ to\ \theta)\ where\ \theta=\pi+\phi$, we can determine the average voltage at the output by using Eq. (3.5).

$$V_{O\ (avg)}=\frac{1}{2\pi}\int_0^{\theta}V_m\sin(\omega t)\cdot d(\omega t)$$

$$V_{O\ (avg)}=\frac{V_m}{\pi}\left(\frac{1-\cos\theta}{2}\right) \tag{3.5}$$

The average value of the current is determined by Eq. (3.3) as shown below.

$$I_{O\ (avg)}=\frac{V_{O\ (avg)}}{R}$$

The average power of the rectifier is determined by the multiplication of voltage and current. Average output power of the half-wave uncontrolled rectifier with inductive load is determined by Eq. (3.6).

$$P_{O\ (avg)}=V_{O\ (avg)}\times I_{O\ (avg)}$$

$$P_{O\ (avg)}=\frac{V_m{}^2}{\pi^2R}\left(\frac{1-\cos\theta}{2}\right)^2 \tag{3.6}$$

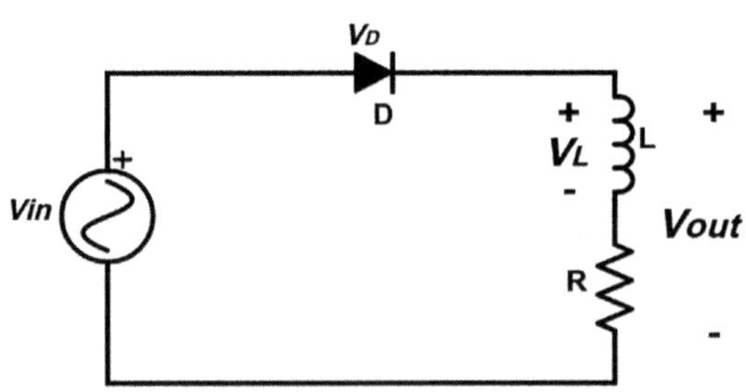

FIGURE 3.15 Half-wave uncontrolled rectifier with an inductive load.

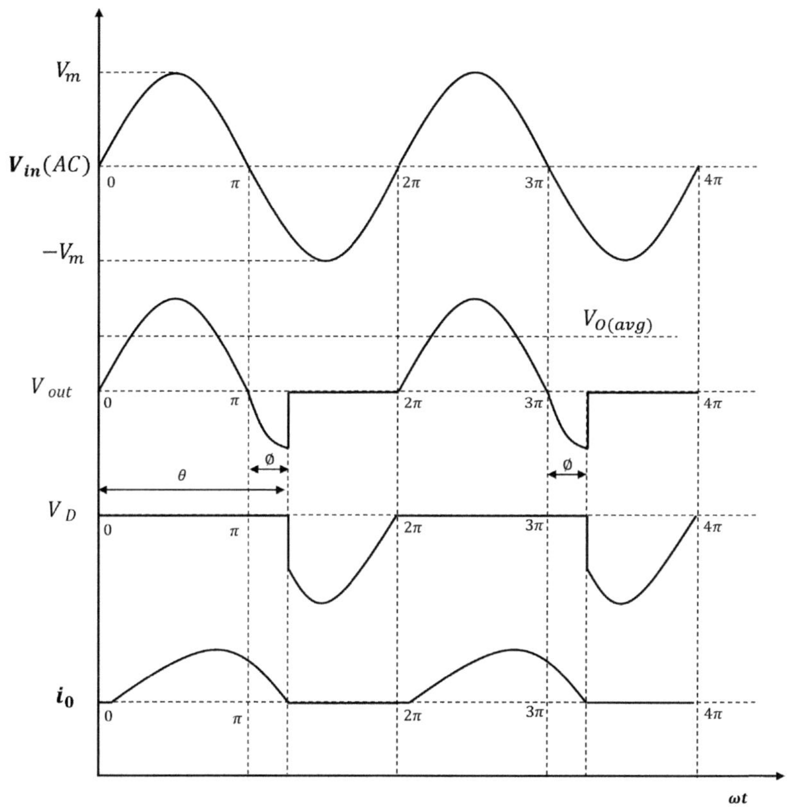

FIGURE 3.16 Waveforms of a half-wave uncontrolled rectifier with an inductive load.

3.3.3 Half-wave uncontrolled rectifier with inductive load and freewheeling diode

To overcome the impact of the inductive load at the output voltage, a freewheeling diode (FWD) is used in parallel to the load that freewheels the energy stored in the inductor during the negative cycle of the input voltage as shown in Fig. 3.17. From Fig. 3.18 it can be seen that during the positive half cycle the voltage is dropped across the load, whereas for the negative half cycle the diode blocks all the input voltage across itself. The output current is the combination of the currents i_{d1} and i_{d2}. i_{d1} is the current through the diode when it is forward biased in positive half cycle, i_{d2} is the current through the freewheeling diode when the diode is reverse biased, and the energy stored in the inductor dissipates through the FWD.

As the waveform of output voltage after using the freewheeling diode is the same as that of the half-wave rectifier with a resistive load, so the equation is used to determine the average output voltage. Because of the

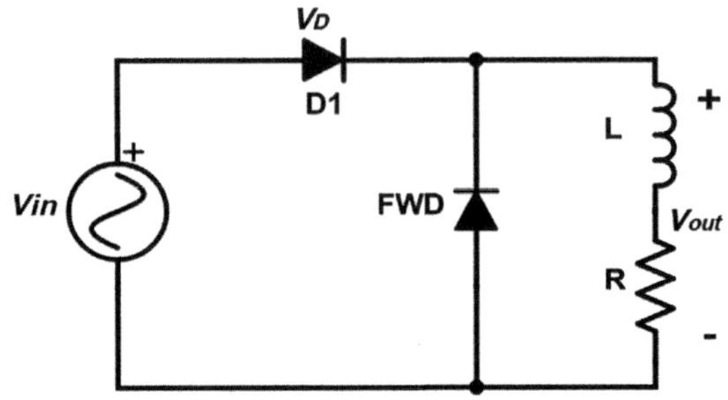

FIGURE 3.17 Half-wave uncontrolled rectifier with an inductive load and a freewheeling diode.

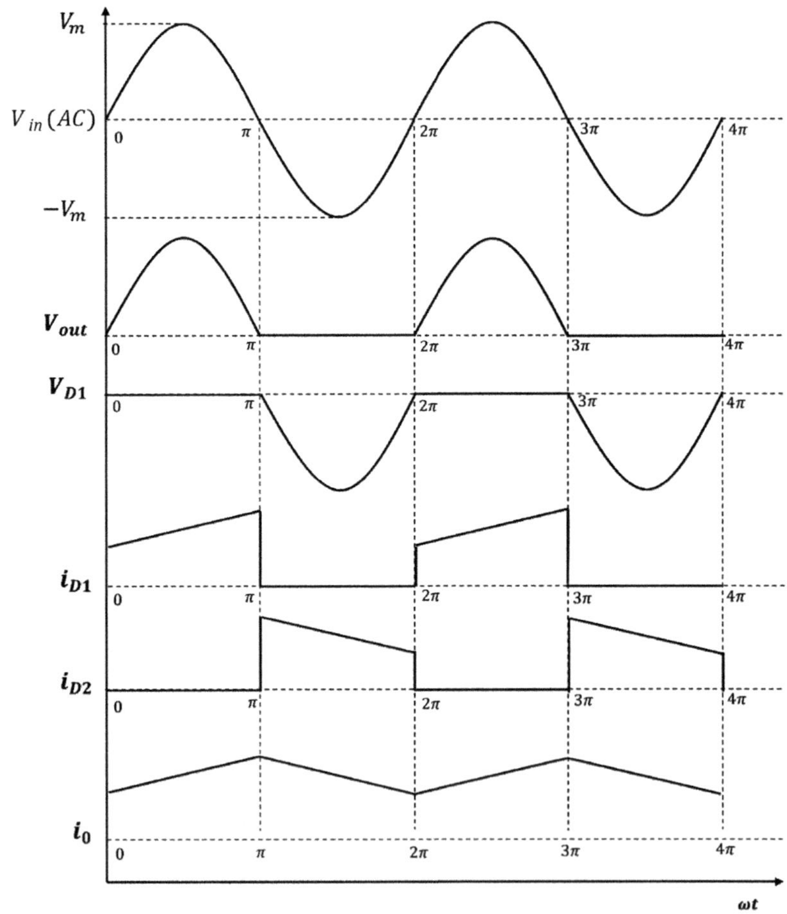

FIGURE 3.18 Output waveforms of a half-wave uncontrolled rectifier with an inductive load and a freewheeling diode.

inductive load, the current waveform is smoother than the rectifier with a resistive load.

3.3.4 Half-wave controlled rectifier with resistive load

On replacing the diode in an uncontrolled rectifier with a thyristor, a controlled rectifier is formed, as shown in Fig. 3.19. During the positive half cycle, the thyristor is forward biased but because of the absence of the gate signal it remains in forward blocking mode. As soon as the thyristor is turned ON by applying the gate signal, a thyristor comes in forward conduction mode and start conducting. From the waveform of V_{AK} in Fig. 3.20, we can see that the voltage is dropped across the thyristor until the gate signal is applied and after the application of the gate signal, the voltage starts appearing across the load. During the negative half cycle, the thyristor becomes reverse biased and drops all the input voltage across itself irrespective of the presence of the gate signal.

The average output voltage can be determined by calculating the area under the one complete cycle using the following equation.

$$V_{O\ (avg)} = \frac{1}{2\pi}\int_0^{2\pi} V_m \sin(\omega t)\cdot d(\omega t)$$

As $V_{Out} = 0$ for the intervals$(0 \le \omega t \le \alpha, \pi \le \omega t \le 2\pi)$, so the above equation becomes

$$V_{O\ (avg)} = \frac{1}{2\pi}\int_\alpha^{\pi} V_m \sin(\omega t)\cdot d(\omega t)$$

$$V_{O\ (avg)} = \frac{V_m}{\pi}\left(\frac{1+\cos\alpha}{2}\right) \tag{3.7}$$

Eq. (3.7) is used to determine the average voltage at the output of the controlled half-wave rectifier with resistive load. The average value of the current is determined by the following relation.

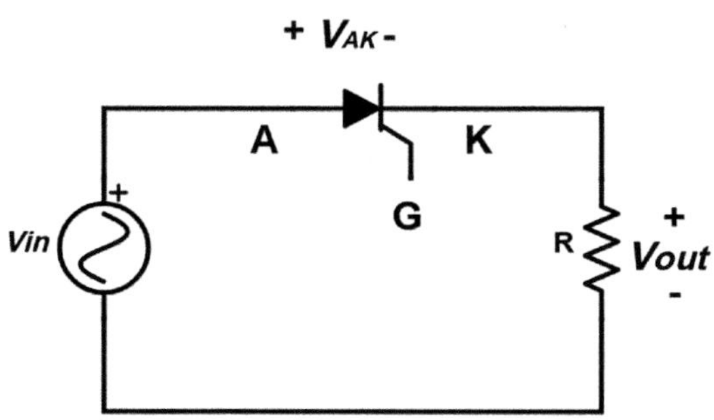

FIGURE 3.19 Half-wave controlled rectifier with a resistive load.

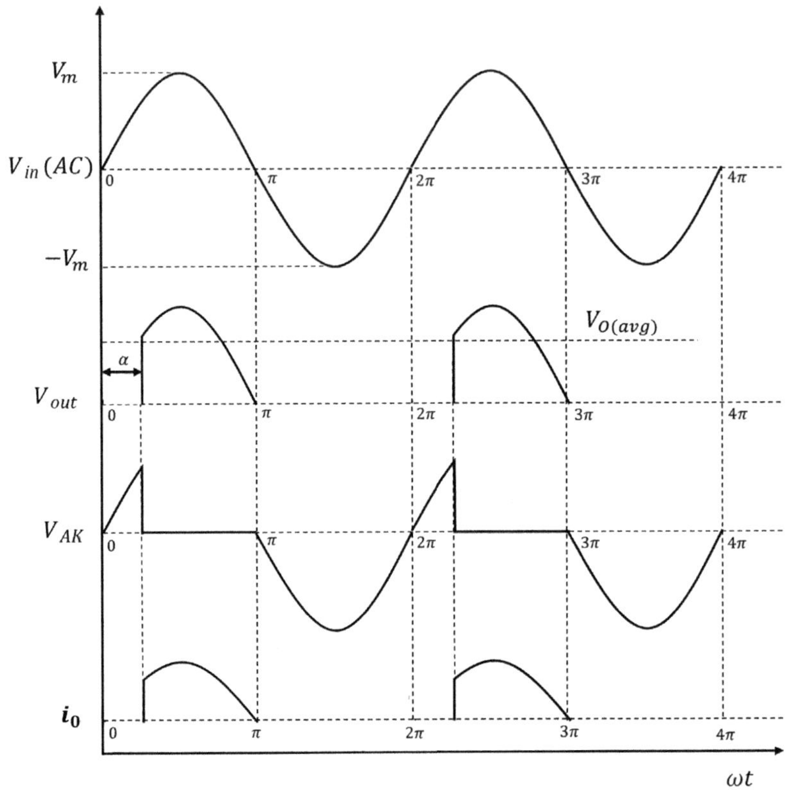

FIGURE 3.20 Output waveforms of a half-wave controlled rectifier with a resistive load.

$$I_{O\ (avg)} = \frac{V_{O\ (avg)}}{R}$$

The average power of the rectifier is determined by the multiplication of voltage and current. Average output power of the half-wave controlled rectifier with resistive load is determined by Eq. (3.8).

$$P_{O\ (avg)} = V_{O\ (avg)} \times I_{O\ (avg)}$$

$$P_{O\ (avg)} = \frac{{V_m}^2}{\pi^2 R}\left(\frac{1+\cos\alpha}{2}\right)^2 \tag{3.8}$$

Example 3.2

In a micro hydro power plant, the output power is to be dumped on a DC link because of the intermittent nature of the flow. To convert the AC output to DC, a half-wave controlled rectifier is used. The input voltage from the micro hydro system is $V_{in} = 170\sin(\omega t)$, If the load resistance is 100 Ω, calculate the output power for the following firing angles. Also interpret your results.

(a) $\alpha = 0^0$
(b) $\alpha = 45^0$
(c) $\alpha = 90^0$
(d) $\alpha = 135^0$
(e) $\alpha = 180^0$

Solution

To calculate the output power of the half-wave controlled rectifier we use Eq. (3.8)

$$P_{O\ (avg)} = \frac{V_m^{\ 2}}{\pi^2 R}\left(\frac{1+\cos\alpha}{2}\right)^2$$

(a) For $\alpha = 0^0$

$$P_{O\ (avg)} = \frac{170^2}{\pi^2 100}\left(\frac{1+\cos 0}{2}\right)^2$$

$$P_{O\ (avg)} = 29.3W$$

(b) For $\alpha = 45^0$

$$P_{O\ (avg)} = \frac{170^2}{\pi^2 100}\left(\frac{1+\cos 45}{2}\right)^2$$

$$P_{O\ (avg)} = 21.3W$$

(c) For $\alpha = 90^0$

$$P_{O\ (avg)} = \frac{170^2}{\pi^2 100}\left(\frac{1+\cos 90}{2}\right)^2$$

$$P_{O\ (avg)} = 7.32W$$

(d) For $\alpha = 135^0$

$$P_{O\ (avg)} = \frac{170^2}{\pi^2 100}\left(\frac{1+\cos 135}{2}\right)^2$$

$$P_{O\ (avg)} = 0.62W$$

(e) For $\alpha = 180^0$

$$P_{O\ (avg)} = \frac{170^2}{\pi^2 100}\left(\frac{1+\cos 180}{2}\right)^2$$

$$P_{O\,(avg)} = 0W$$

From the above calculations of power at different firing angles we can estimate that by increasing the firing angle the power at the output is reduced.

3.3.5 Half-wave controlled rectifier with an inductive load

A controlled rectifier is the same as the uncontrolled rectifier except that the diode is replaced by the thyristor, as shown in Fig. 3.21. During the positive half cycle, the thyristor is forward biased but because of the absence of the gate signal it remains in forward blocking mode. As soon as the thyristor is turned ON by applying the gate signal at angle α, the thyristor comes into forward conduction mode and starts conducting. From the waveform of anode to cathode voltage (V_{AK}) in Fig. 3.22, we can see that the voltage is dropped across the thyristor until the gate signal is applied and after the application of the gate signal, the voltage starts appearing across the load. From π onward, the input voltage starts increasing in the reverse direction making the diode reverse biased, but the current through the inductor starts decaying and voltage across it builds up with opposite polarity until reaching the angle ϕ. From ϕ to 2π, the thyristor is reverse biased and all the input voltage is dropped across the thyristor and zero voltage across the load. After that the cycle repeats itself. Because of the inductive load, the output current lags behind the output voltage [4].

The average output voltage can be determined by calculating the area under the one complete cycle using the equation.

$$V_{O\,(avg)} = \frac{1}{2\pi}\int_0^{2\pi} V_m \sin(\omega t)\cdot d(\omega t)$$

As $V_{Out} = 0$ for the intervals $(0 \le \omega t \le \alpha, \theta \le \omega t \le 2\pi)$, so the above equation becomes.

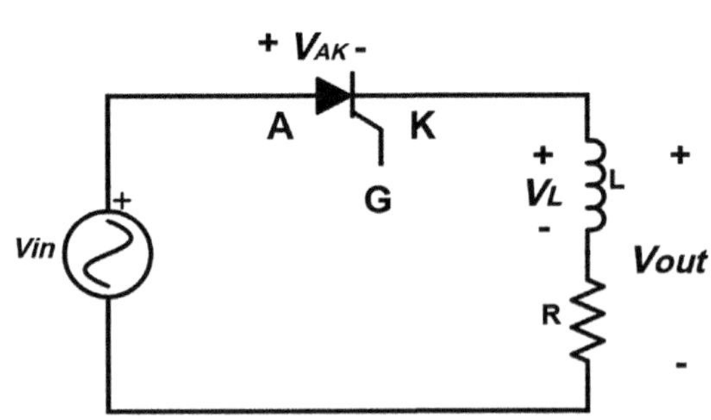

FIGURE 3.21 Half-wave controlled rectifier with an inductive load.

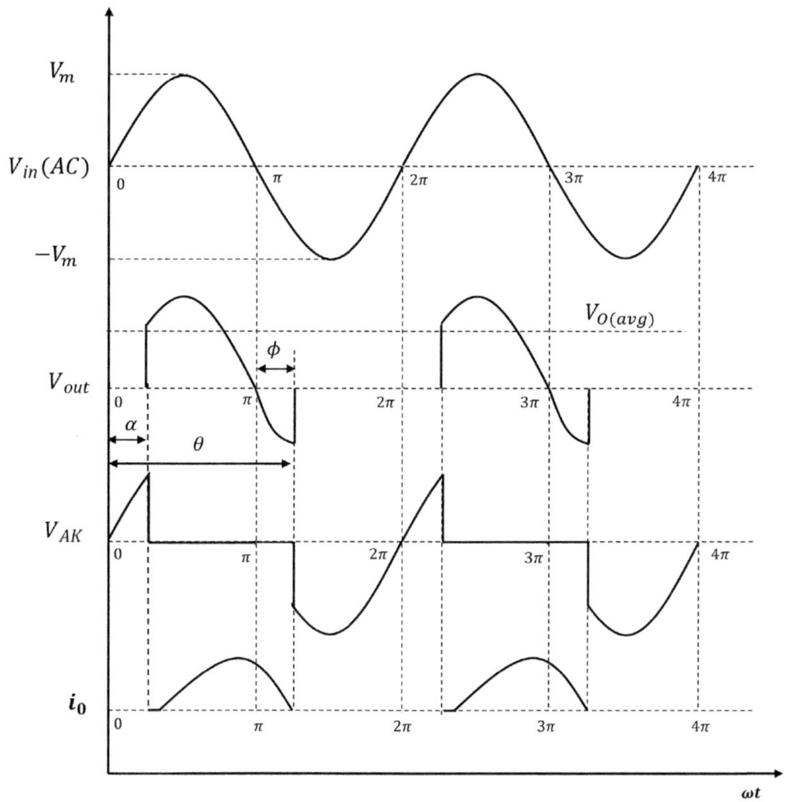

FIGURE 3.22 Output waveforms of a half-wave controlled rectifier with an inductive load.

$$V_{O\ (avg)} = \frac{1}{2\pi}\int_{\alpha}^{\theta} V_m \sin(\omega t)\cdot d(\omega t)$$

$$V_{O\ (avg)} = \frac{V_m}{\pi}\left(\frac{\cos\alpha - \cos\theta}{2}\right) \tag{3.9}$$

Eq. (3.9) gives the average voltage at the output of the half-wave controlled rectifier with inductive load. The average value of the current is determined by dividing the average voltage to load resistance, as shown in the following relation.

$$I_{O\ (avg)} = \frac{V_{O\ (avg)}}{R}$$

The average power of the rectifier is determined by the multiplication of voltage and current. Eq. (3.10) gives the average output power of the half-wave controlled rectifier with inductive load.

$$P_{O\ (avg)} = V_{O\ (avg)} \times I_{O\ (avg)}$$

$$P_{O\ (avg)} = \frac{V_m{}^2}{\pi^2 R}\left(\frac{\cos\alpha - \cos\theta}{2}\right)^2 \tag{3.10}$$

3.3.6 Half-wave controlled rectifier with inductive load and a freewheeling diode

To overcome the impact of the inductive load at the output voltage, a freewheeling diode is used in parallel to the load that freewheels the energy stored in the inductor during the negative cycle of the input voltage as shown in Fig. 3.23. From Fig. 3.24 it can be seen that during the positive half cycle after the application of the gate signal α the voltage is dropped across the load, whereas for a negative half cycle the thyristor blocks all the input voltage across itself. The output current is the combination of the thyristor current when it is triggered in positive half cycle and the freewheeling diode current when the thyristor is reverse biased and the energy stored in the inductor dissipates through the FWD.

As the waveform of output voltage after using the freewheeling diode is the same as that of the half-wave rectifier with resistive load, so the equation is used to determine the average output voltage. Because of the inductive load, the current waveform is smoother than the rectifier with resistive load. i_{SCR} is the current flowing through the thyristor when it is in the forward conduction mode. Its behavior is increasing as the inductive load stores energy gradually in its magnetic field. i_{FWD} is the freewheeling diode current during the reverse biasing of the thyristor and its behavior is decreasing as the energy stored in the inductor dissipates gradually through the FWD.

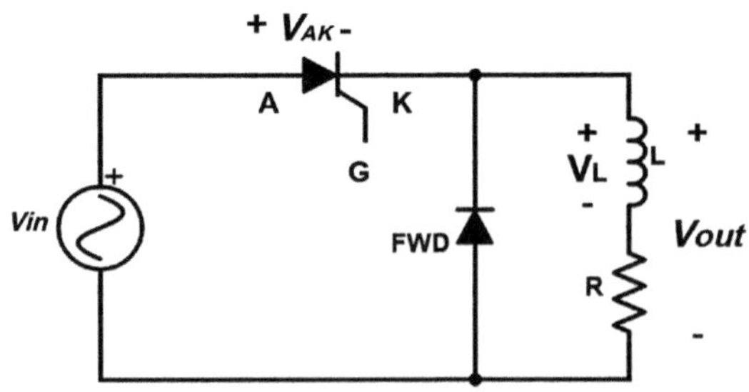

FIGURE 3.23 Half-wave controlled rectifier with an inductive load and a freewheeling diode.

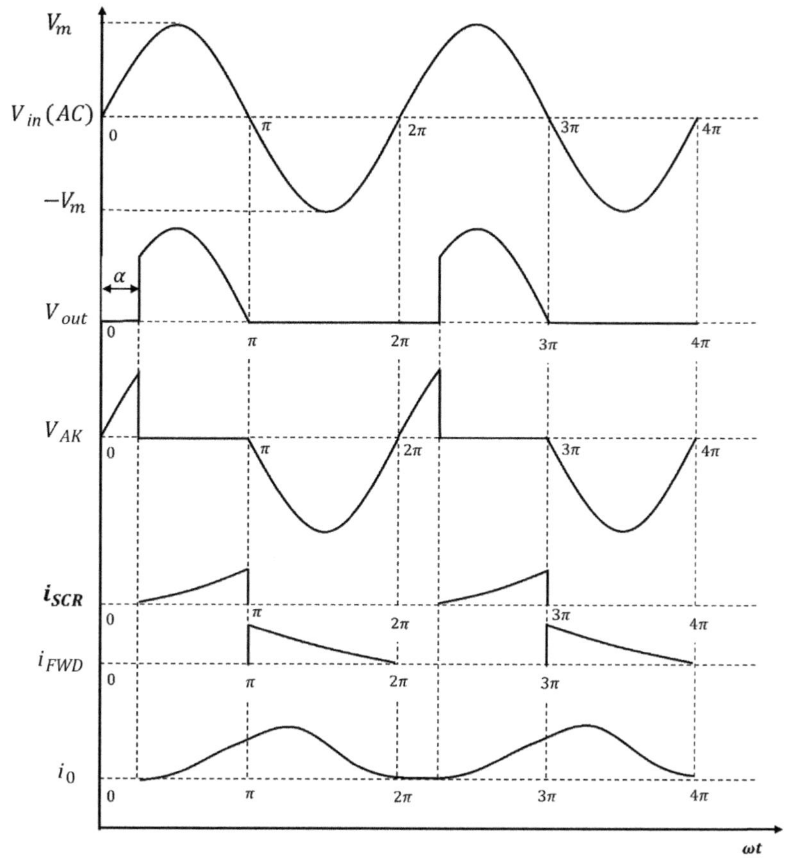

FIGURE 3.24 Output waveforms of a half-wave controlled rectifier with an inductive load and a freewheeling diode.

3.4 Converters (DC–DC converters)

3.4.1 Buck converters

We assume that the ON time of the switching element S is DT and OFF time is $D'T$. Where D is the duty cycle when the switching element is ON, D' is the duty cycle when the switching element is OFF, and T is the time period of the switching element which is the sum of both DT and $D'T$, as shown in Eq. (3.11), which can also be written as Eq. (3.12). Fig. 3.25 represents the single pole double throw switch based diagram of DC-DC buck converter. A single pole double throw switch is realized by replacing it with an active switch like a thyristor and a passive switch like a diode. Fig. 3.26 shows the DC-DC buck converter circuit after the realization of the single pole double throw switch.

$$T = DT + D'T \tag{3.11}$$

$$T = T_{ON} + T_{OFF} \tag{3.12}$$

Sum of the duty cycle during T_{ON} and T_{OFF} is 1, that is, $D + D' = 1$.

When the switch is in position 1 or is ON it behaves like a short circuit and the current exponentially builds up through the inductor and the load. The inductor voltage V_L reaches the difference of input voltage V_{in} and the output voltage V_{out}. The reverse polarity of the voltage across the diode reverse biases the diode and it behaves like an open circuit, as shown in Fig. 3.27. To determine the relations, apply KVL on the input loop during

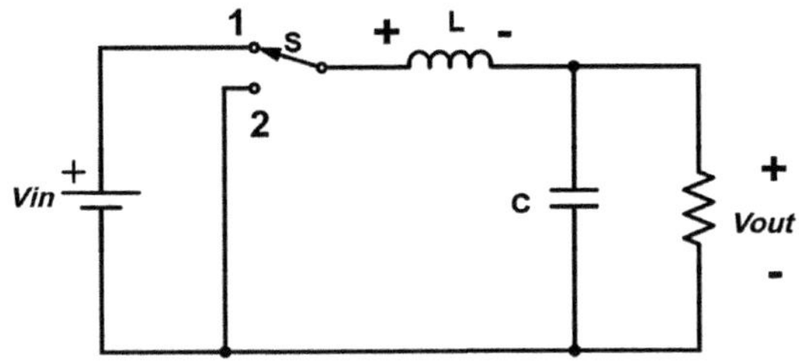

FIGURE 3.25 DC–DC buck converter with single-pole double-throw switch.

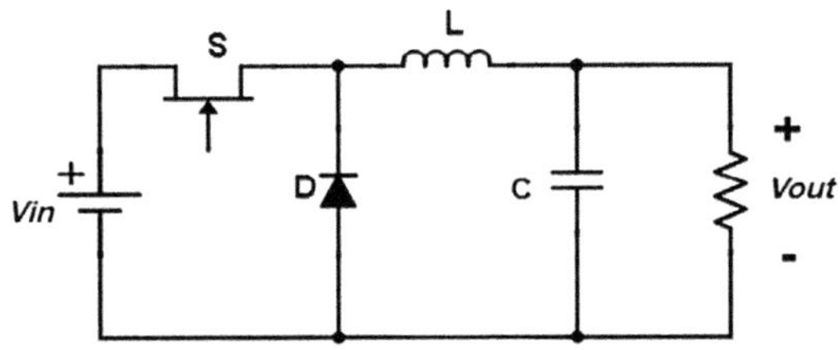

FIGURE 3.26 DC–DC buck converter with switch realization.

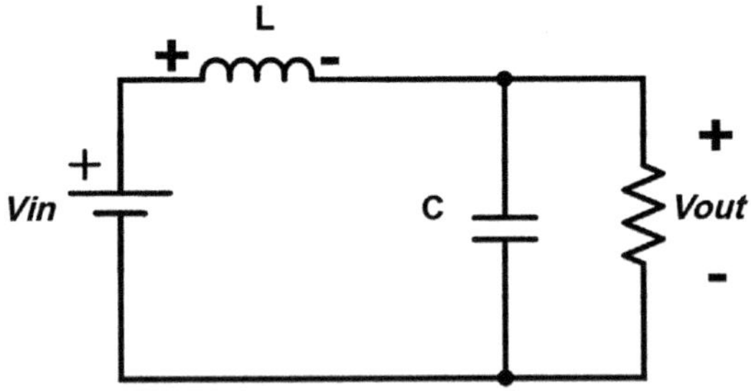

FIGURE 3.27 DC–DC buck converter when the switch S is ON.

$$V_{in} - V_L - V_{out} = 0$$

$$V_L = V_{in} - V_{out} \tag{3.13}$$

T_{ON}. Eq. (3.13) is used to determine the inductor voltage when the switch is ON [5].

When the switch is in position 2 or is OFF it behaves like an open circuit, the voltage source is disconnected and the voltage that was developed across the inductor during T_{ON} is now equal to the V_{out}, as shown in Fig. 3.28. Apply KVL at the output loop during T_{OFF}. Eq. (3.14) is used to determine the inductor voltage when the switch is OFF [6].

$$-V_L - V_{out} = 0$$

$$V_L = -V_{out} \tag{3.14}$$

The voltage across the inductor is averaged by dividing the area under the curve shown in Fig. 3.29 to the switching time period T.

$$\langle V_L \rangle = \frac{(V_{in} - V_{out})(DT) + (-V_{out})(D'T)}{T}$$

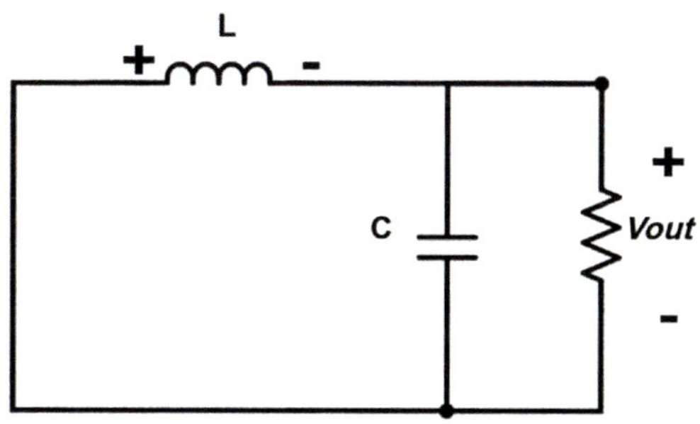

FIGURE 3.28 DC–DC buck converter when the switch S is OFF.

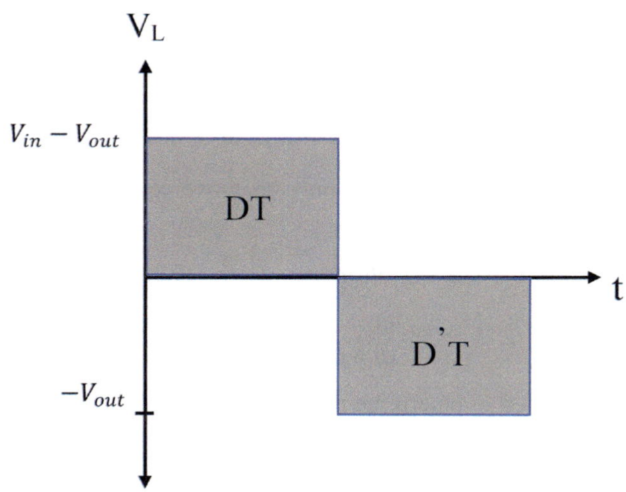

FIGURE 3.29 Inductor voltage in DC–DC buck converter.

As the average voltage across the inductor is zero, so

$$0 = \frac{T[D(V_{in} - V_{out}) - D'V_{out}]}{T}$$

$$0 = DV_{in} - DV_{out} - D'V_{out}$$

$$0 = DV_{in} - V_{out}(D + D')$$

$$V_{out} = DV_{in} \tag{3.15}$$

This relation of Eq. (3.15) is used to determine the required duty cycle for your own output voltage requirements. For example, if your solar photovoltaic system is of 24 V but your battery system is of 12 V, you will have to determine the duty cycle of your converter using the above derived Eq. (3.15).

To determine the relation for the inductor, we need to calculate the rate of change of current through the inductor during T_{ON} and T_{OFF}. The voltage across the inductor is given by Eq. (3.16).

$$V_L = L.\frac{di}{dt_{(DT)}} \tag{3.16}$$

By Eq. (3.13) during T_{ON}, $V_L = V_{in} - V_{out}$, so put it in Eq. (3.16).

$$L\frac{di}{dt_{(DT)}} = V_{in} - V_{out}$$

$$\frac{di}{dt_{(DT)}} = \frac{V_{in} - V_{out}}{L} \tag{3.17}$$

Eq. (3.17) is used to determine the change in current through the inductor when the switch is ON. Similarly during T_{OFF} Eq. (3.16) can be written as follow:

$$V_L = L.\frac{di}{dt_{(D'T)}}$$

Placing Eq. (3.14), $V_L = -V_{out}$, into the above equation.

$$L\frac{di}{dt_{(D'T)}} = -V_{out}$$

$$\frac{di}{dt_{(D'T)}} = \frac{-V_{out}}{L} \tag{3.18}$$

Eq. (3.18) is used to determine the change in current through the inductor when the switch is OFF. The graph of the inductor current can be drawn by using equations for DT and $D'T$. Eqs. (3.17) and (3.18) represent the slope of the inductor current that can be used to find out the value of the inductor to be used in the buck converter by multiplying the slope of the inductor current to the subinterval as shown in Fig. 3.30 and calculated in Eq. (3.19).

Example 3.3

A battery connected solar PV system of 24 V input voltage has to charge the battery system of 12 V for which a buck converter of inductor 8.2 H is used. Calculate

(a) The duty cycle of the boost converter.
(b) Inductor ripple current.

Solution

(a) We can find the duty cycle of the buck converter by using

$$V_{out} = DV_{in}$$

$$12 = D.24$$

$$D = 0.5$$

(b) The inductor ripple current can be calculate by modifying Eq. (3.19)

$$L = \frac{V_{in} - V_{out}}{2\Delta i_L}.DT$$

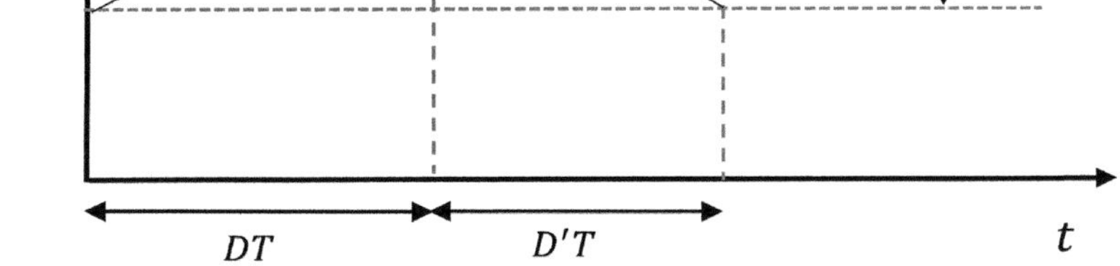

FIGURE 3.30 Inductor current in DC–DC buck converter.

$$\text{Change in } i_L = (\text{slope})(\text{length of sub} - \text{interval})$$

$$2\Delta i_L = \frac{V_{in} - V_{out}}{L}.DT$$

$$L = \frac{V_{in} - V_{out}}{2\Delta i_L}.DT \tag{3.19}$$

$$\Delta i_L = \frac{V_{in} - V_{out}}{2L}.DT$$

$$\Delta i_L = \frac{24 - 12}{2(8.2)}.(0.5)$$

$$\Delta i_L = 365\ mA$$

3.4.2 Boost converters

We assume that the ON time of the switching element S is DT and OFF time is $D'T$.Where D is the duty cycle when the switching element is ON, D' is the duty cycle when the switching element is OFF, and T is the time period of the switching element.

$$T = DT + D'T$$

$$T = T_{ON} + T_{OFF}$$

The sum of the duty cycle during T_{ON} and T_{OFF} is 1, i.e., $D + D' = 1$. Fig. 3.31 shows the circuit diagram of the DC–DC boost converter with an single-pole double-throw (SPDT) switch. After realization of the SPDT switch, Fig. 3.32 shows the boost converter circuit.

When the switch is in position 1 or is ON, the inductor is disconnected from the output and connected to the input only because of the reverse biasing of the diode. The inductor voltage immediately jumps to the input voltage and the current builds up linearly storing energy in the magnetic field of the inductor. By applying Kirchhoff's Voltage Law (KVL) at the input loop of Fig. 3.33, we can find the inductor voltage using Eq. (3.20).

When the switch is in position 2 or is OFF, the current starts decaying, which reverses the polarity of the voltage across the inductor. This reverse polarity forward biases the diode, which behaves like a short circuit. Now the inductor is connected to both the input and the output. The inductor

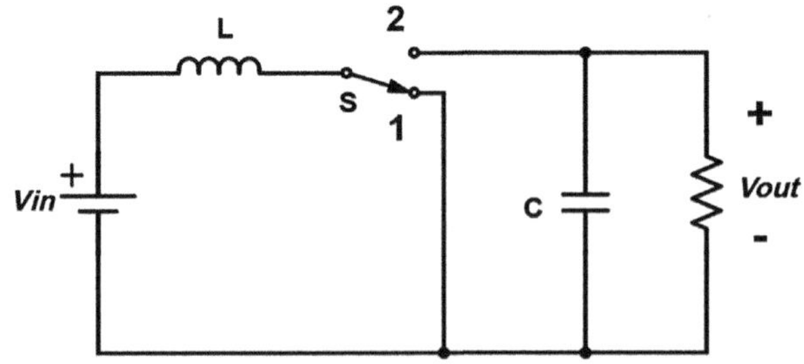

FIGURE 3.31 DC–DC boost converter with single-pole double-throw switch.

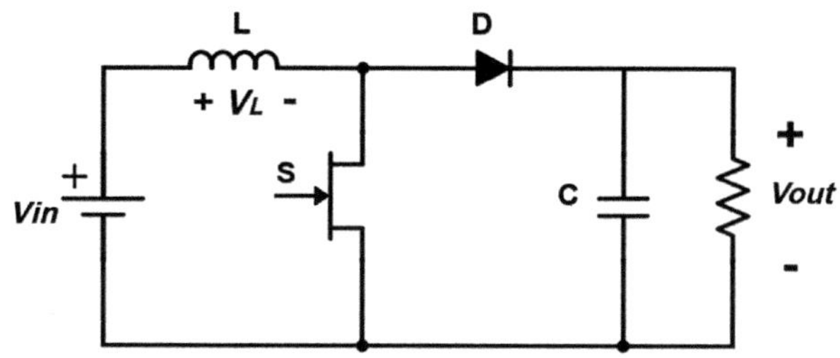

FIGURE 3.32 DC–DC boost converter with switch realization.

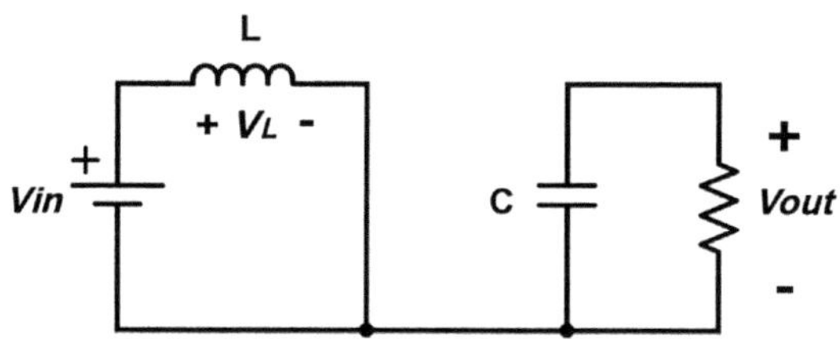

FIGURE 3.33 DC–DC boost converter when switch S is ON.

$$V_{in} - V_L = 0$$

$$V_L = V_{in} \tag{3.20}$$

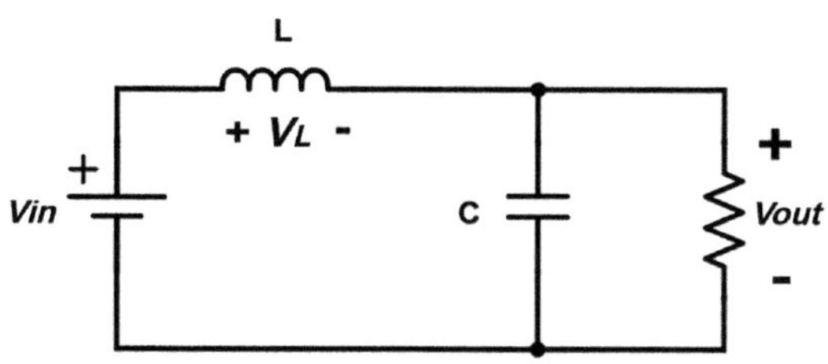

FIGURE 3.34 DC–DC boost converter when switch S is OFF.

$$V_{in} - V_L - V_{out} = 0$$

$$V_L = V_{in} - V_{out} \tag{3.21}$$

current starts to flow through the diode and the load. The inductor voltage can be found by applying KVL to the circuit shown in Fig. 3.34.

Eqs. (3.20) and (3.21) found during both the intervals can graphically be represented in Fig. 3.35. The voltage across the inductor is averaged by dividing the area under the curve in Fig. 3.35 to the switching time period T.

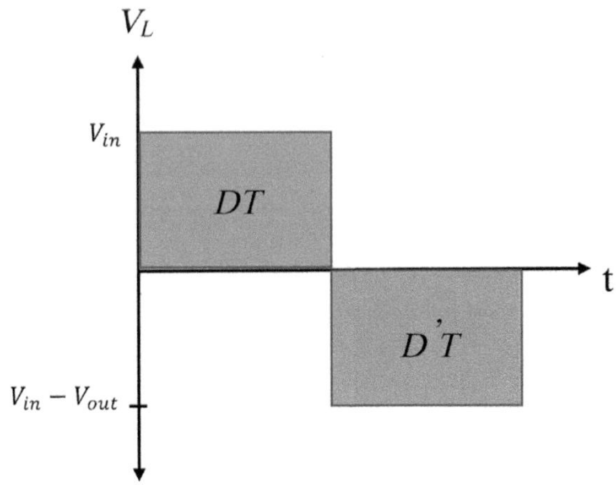

FIGURE 3.35 Inductor voltage in DC–DC boost converter.

$$\langle V_L \rangle = \frac{(V_{in})(DT) + (V_{in} - V_{out})(D'T)}{T}$$

As the average voltage across the inductor is zero, so

$$0 = V_{in}D + V_{in}D' - V_{out}D'$$

$$0 = V_{in}(D + D') - V_{out}D'$$

$$V_{out} = \frac{V_{in}}{1 - D} \tag{3.22}$$

This relation of Eq. (3.22) is used to determine the required duty cycle for your own output voltage requirements. For example, if your solar photovoltaic system is of 12 V but your battery system is of 24 V, you will have to determine the duty cycle of your converter using the above derived equation. For different values of duty cycle D in the range 0.1 to 0.9, the output voltage will be greater than the input voltage.

To determine the relation for the inductor, we need to calculate the rate of change of current through the inductor during T_{ON} and T_{OFF}. The voltage across the inductor is given by the Eq. (3.16) shown below.

$$V_L = L.\frac{di}{dt_{(DT)}}$$

By Eq. (3.20) during T_{ON}, $V_L = V_{in}$, so the above equation becomes

$$L\frac{di}{dt_{(DT)}} = V_{in}$$

$$\frac{di}{dt_{(DT)}} = \frac{V_{in}}{L} \tag{3.23}$$

Eq. (3.23) is used to determine the change in current through the inductor when the switch is ON. Similarly during T_{OFF} Eq. (3.16) becomes:

$$V_L = L.\frac{di}{dt_{(D'T)}}$$

By Eq. (3.21) during T_{OFF}, $V_L = V_{in} - V_{out}$, so the above equation becomes

$$L\frac{di}{dt_{(D'T)}} = V_{in} - V_{out}$$

$$\frac{di}{dt_{(D'T)}} = \frac{V_{in} - V_{out}}{L} \tag{3.24}$$

The graph of the inductor current can be drawn by using equations for DT and $D'T$. Eqs. (3.23) and (3.24) represent the slope of the inductor current that can be used to find out the value of the inductor to be used in the boost converter by multiplying the slope of the inductor current to the subinterval as shown in Fig. 3.36.

As seen in Fig. 3.32, we can say that the inductor current is the input current $i_i = i_L$ that can be averaged as follows.

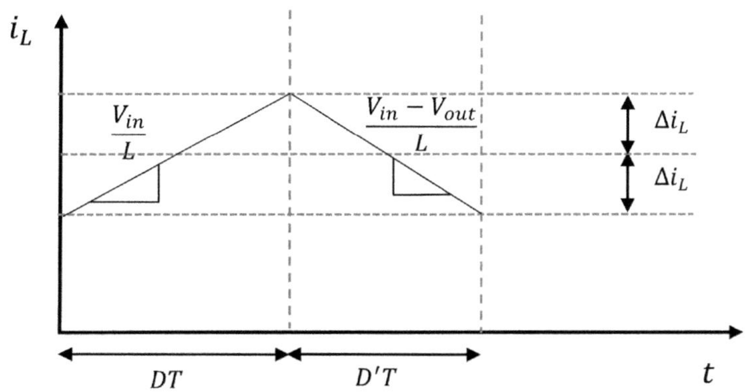

FIGURE 3.36 Inductor current in DC–DC boost converter.

$$\text{Change in } i_L = (\text{slope})(\text{length of sub} - \text{interval})$$

$$2\Delta i_L = \frac{V_{in}}{L} \cdot DT$$

$$L = \frac{V_{in}}{2\Delta i_L} \cdot DT \tag{3.25}$$

$$P_{in} = P_{out}$$

$$V_i.\langle i_i \rangle = \frac{\langle V_{out} \rangle^2}{R}$$

$$\langle i_i \rangle = \frac{\langle V_{out} \rangle^2}{V_i.R}$$

So by inserting the value of the inductor current in Eq. (3.25) it takes the form

$$L = \frac{(V_{in})^2 \cdot R}{2\langle V_{out} \rangle^2} \cdot DT$$

Example 3.4

A battery connected solar PV system of 12 V input voltage has to charge the battery system of 24 V for which a boost converter is used. To get the voltage of the PV system equal to the battery voltage:

(a) what should be the duty cycle of the boost converter?
(b) If the load of 5 $k\ \Omega$ is connected, calculate the inductor value of the boost converter.

Solution

(a) The duty cycle of the boost converter is given by

$$V_{out} = \frac{V_{in}}{1 - D}$$

$$24V = \frac{12V}{1 - D}$$

$$D = 0.5$$

(b) The value of the inductor of the boost converter is given by

$$L = \frac{(V_{in})^2 \cdot R}{2\langle V_{out} \rangle^2} \cdot DT$$

$$L = \frac{(12)^2 \cdot 5k}{2(24)^2} \times (0.5)$$

$$L = 312H$$

3.4.3 Buck-Boost converters

We assume that the ON time of the switching element S is DT and OFF time is $D'T$. Where D is the duty cycle when the switching element is ON, D' is the duty cycle when the switching element is OFF, and T is the time period of the switching element.

$$T = DT + D'T$$

$$T = T_{ON} + T_{OFF}$$

Sum of the duty cycle during T_{ON} and T_{OFF} is 1, that is, $D + D' = 1$

Fig. 3.37 shows the circuit of a buck-boost converter. SPDT switch has been used as a switching element. At first throw, it connects the inductor to the input and charges it, whereas at the second throw it connects the inductor to the output and transfers the stored energy. SPDT can be replaced by two SPST switches where one should be closed at a time and the other should be open. As we are dealing with power electronic devices, accuracy and perfection are required. So, a single-pole double throw switch can be realized by a controlled and an uncontrolled switch, as shown in Fig. 3.38. The controlled switch is turned on and off by a gate signal, whereas the controlled switch is turned on and off by the voltage polarity across it.

When the switch S is in position 1 or is ON it behaves like a short circuit, the inductor voltage V_L abruptly reaches the input voltage V_{in} and the current linearly charges the inductor with time. The reverse polarity of the voltage

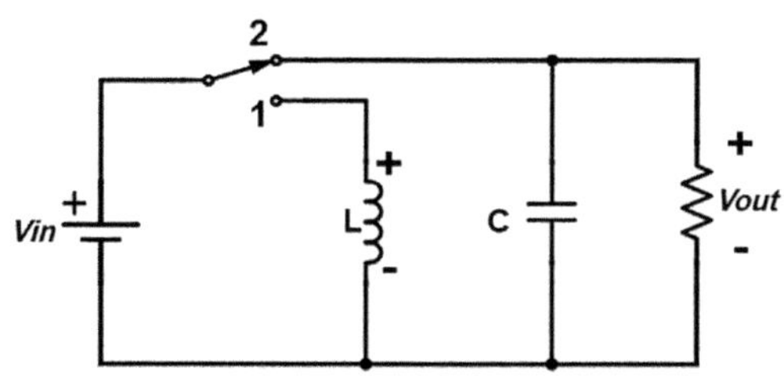

FIGURE 3.37 DC–DC buck-boost converter with single-pole double-throw switch.

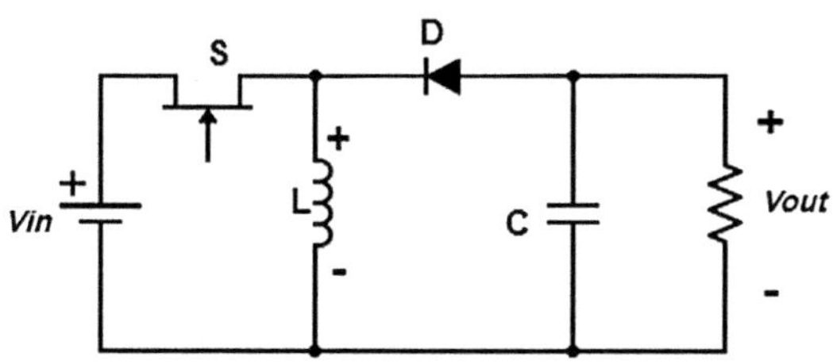

FIGURE 3.38 DC–DC buck-boost converter with switch realization.

across the diode reverse biases the diode and it behaves like an open circuit, as shown in Fig. 3.39. To determine the relations, apply KVL on the input loop during turn ON time and we get Eq. (3.26).

$$V_{in} - V_L = 0$$

$$V_L = V_{in} \tag{3.26}$$

When the switch S is in position 2 or is OFF it behaves like an open circuit, the voltage source is disconnected, the inductor starts to release its stored energy with polarity reversed, and the voltage that was developed across the inductor during T_{ON} is now equal to the V_{out}, as shown in Fig. 3.40. Applying KVL at the output loop during T_{OFF} obtains Eq. (3.27) for inductor voltage when the switch is OFF.

$$V_L - V_{out} = 0$$

$$V_L = V_{out} \tag{3.27}$$

The voltage across the inductor is averaged by dividing the area under the curve shown in Fig. 3.41 to the switching time period T.

$$\langle V_L \rangle = \frac{V_{in}(DT) + V_{out}(D'T)}{T}$$

As the average voltage across the inductor is zero,

$$0 = \frac{T(DV_{in} + D'V_{out})}{T}$$

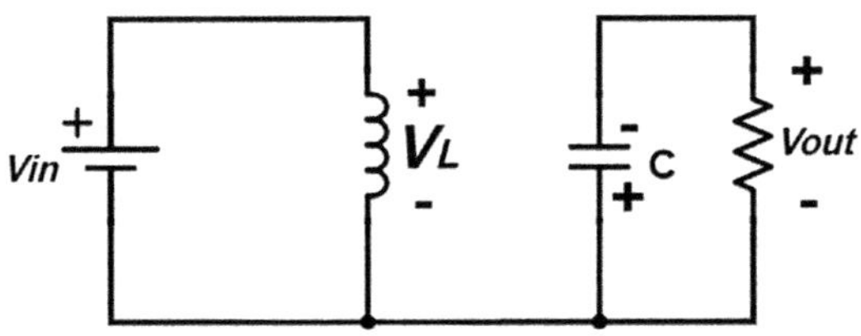

FIGURE 3.39 DC–DC buck-boost converter when switch 'S' is ON.

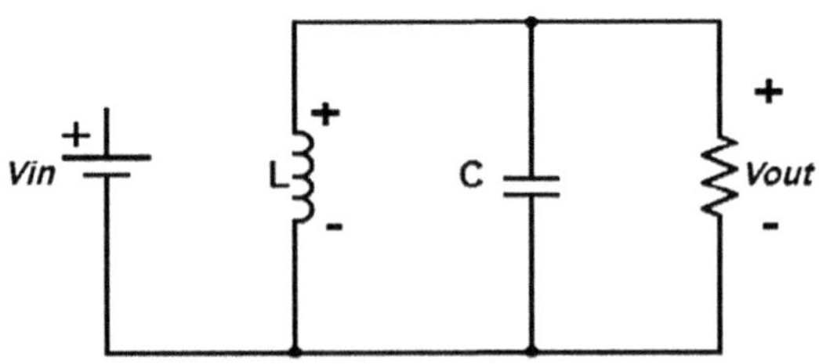

FIGURE 3.40 DC–DC buck-boost converter when switch S is OFF.

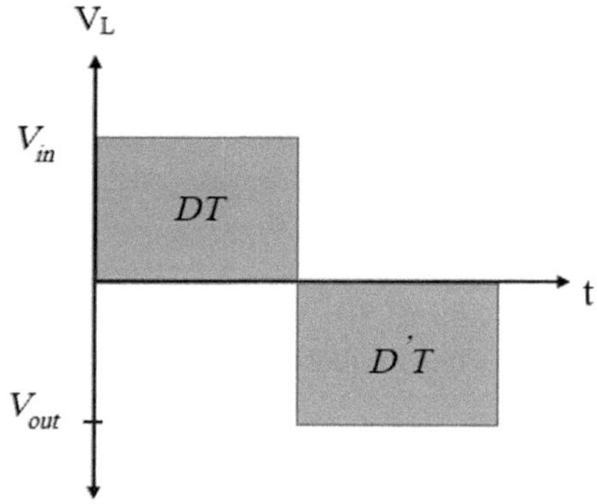

FIGURE 3.41 Inductor voltage in DC–DC buck-boost converter.

$$V_{out} = -\frac{D}{D'}V_{in}$$

$$V_{out} = -\left(\frac{D}{1-D}\right)V_{in} \tag{3.28}$$

Eq. (3.28) is used to determine the required duty cycle for your own output voltage requirements. For example, if your solar photovoltaic system is of 12 V but your battery system is of 24 V, you will have to determine the duty cycle of your converter using the above derived relation in equation.

To determine the relation for the inductor, we need to calculate the rate of change of current through the inductor during T_{ON} and T_{OFF}. The voltage across the inductor is given by Eq. (3.16) as given below.

$$V_L = L \cdot \frac{di}{dt(DT)}$$

By Eq. (3.26), $V_L = V_{in}$, so the above equation becomes:

$$L \cdot \frac{di}{dt(DT)} = V_{in}$$

$$\frac{di}{dt(DT)} = \frac{V_{in}}{L} \tag{3.29}$$

Eq. (3.29) is used to determine the change in current through the inductor when the switch is ON. Similarly during T_{OFF}.

$$V_L = L \cdot \frac{di}{dt(D'T)}$$

By Eq. (3.27), $V_L = V_{out}$, so

$$L \cdot \frac{di}{dt_{(D'T)}} = V_{out}$$

$$\frac{di}{dt_{(D'T)}} = \frac{V_{out}}{L} \tag{3.30}$$

By using Eqs. (3.29) and (3.30) we can draw the graph of the inductor current for DT and $D'T$. These equations represent the slope of the inductor

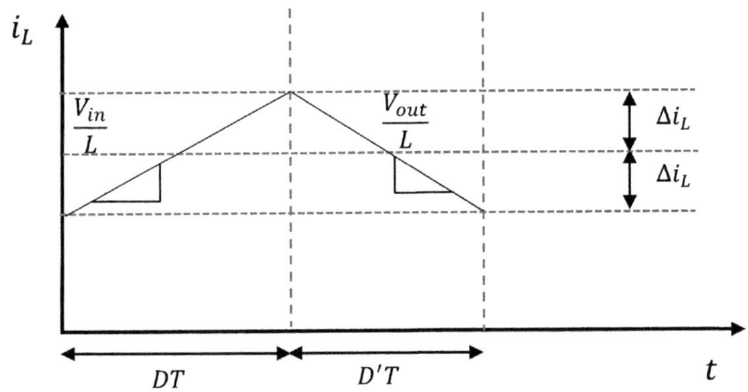

FIGURE 3.42 Inductor current in DC–DC buck-boost converter.

$$\text{Change in } i_L = (\text{slope})(\text{length of sub} - \text{interval})$$

$$2\Delta i_L = \frac{V_{in}}{L} \cdot DT$$

$$L = \frac{V_{in}}{2\Delta i_L} \cdot DT \tag{3.31}$$

current that can be used to find out the value of the inductor to be used in the buck-boost converter, as shown in Fig. 3.42.

3.4.4 Cuk converters

In a Cuk converter, contrary to the buck-boost converter, the boost converter is cascaded by a buck converter. Fig. 3.43 shows the refined form of the Cuk converter with conventional mechanical switches. When the switch 1 is ON, switch 2 is OFF, and vice versa.

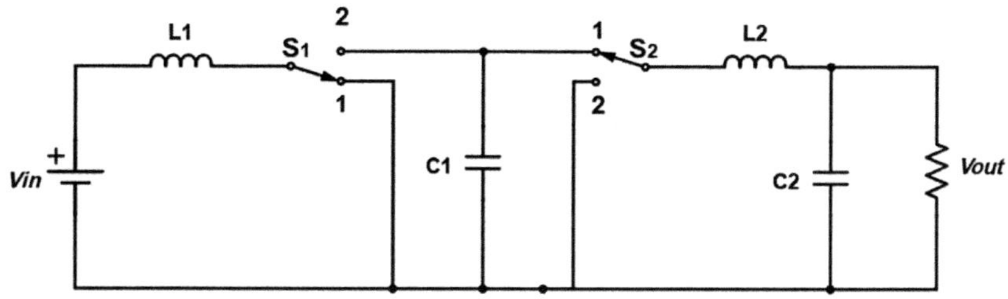

FIGURE 3.43 DC–DC Cuk converter.

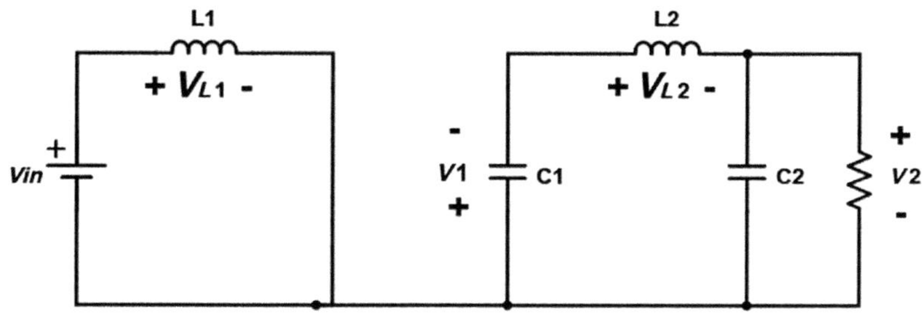

FIGURE 3.44 DC–DC Cuk converter when switch S1 and S2 are in position 1.

When the switch 1 and switch 2 are in position 1, the inductor L_1 is connected to the input. The input voltage is instantaneously developed across L_1, whereas the current linearly builds up and stores energy in the magnetic field of L_1. Capacitor C_1 between the two switches is disconnected from the input and releases its stored energy to the output through the S_2, and L_2 as shown in Fig. 3.44. By applying KVL at the input loop we get the following equations.

$$V_{in} - V_{L1} = 0$$

$$V_{L1} = V_{in} \tag{3.32}$$

Eq. (3.32) gives the voltage across L_1 when the switches are in position 1. Similarly, by applying KVL at the output loop we get Eq. (3.33) that gives the voltage across L_2 when the switches are in position 1.

$$-V_1 - V_{L2} - V_2 = 0$$

$$V_{L2} = -V_1 - V_2 \tag{3.33}$$

When the switch 1 and switch 2 are in position 2, the discharged capacitor is now disconnected from the output and now it is charged by the input voltage and the inductor voltage (V_{L1}). The load is entertained by the V_{L2}. By applying KVL at the input loop of Fig. 3.45 we get Eq. (3.34).

$$V_{in} - V_{L1} - V_1 = 0$$

$$V_{L1} = V_{in} - V_1 \tag{3.34}$$

Eq. (3.34) gives the voltage across L_1 when the switches are in position 2. Similarly, by applying KVL at the output loop of Fig. 3.45 we get Eq. (3.35) that gives the voltage across L_2 when the switches are in position 2.

$$-V_{L2} - V_2 = 0$$

$$V_{L2} = -V_2 \tag{3.35}$$

The voltage across the inductors L_1 and L_2 found during T_{ON} and T_{OFF} can graphically be presented as shown in Fig. 3.46A,B respectively.

The voltage across the inductor L_1 is averaged by dividing the area under the curve shown in Fig. 3.46A to the switching time period T.

$$\langle V_{L1}\rangle = \frac{(V_{in})(DT) + (V_{in} - V_1)(D'T)}{T}$$

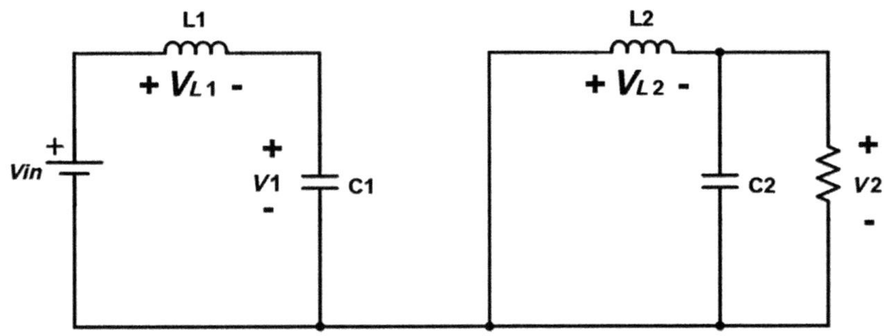

FIGURE 3.45 DC–DC cuk converter when switch S1 and S2 are in position 2.

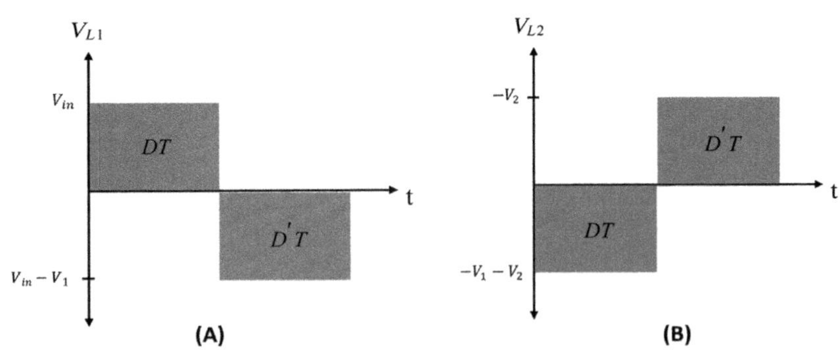

FIGURE 3.46 DC–DC Cuk converter (A) Inductor voltage V_{L1}. (B) Inductor voltage V_{L2}.

$$O = V_{in}D + V_{in}D' - V_1D'$$

$$O = V_{in}(D + D') - V_1D'$$

$$V_1 = \frac{V_{in}}{D'} \tag{3.36}$$

Similarly, the voltage across the inductor L_2 is averaged by dividing the area under the curve in Fig. 3.46B to the switching time period T.

$$\langle V_{L2} \rangle = \frac{(-V_1 - V_2)(DT) + (-V_2)(D'T)}{T}$$

$$O = -V_1D - V_2D - V_2D'$$

$$O = -V_2(D + D') - V_1D$$

$$V_2 = -V_1D$$

By taking V_1 from Eq. (3.36)

$$V_2 = -\frac{V_{in}}{D'}D$$

$$V_2 = -\left(\frac{D}{1 - D}\right)V_{in} \tag{3.37}$$

Eq. (3.37) gives the relation between input voltage and the output voltage depending upon the duty cycle. To determine the relation for the inductor L_1, we need to calculate the rate of change of current through the inductor L_1 during T_{ON} and T_{OFF}, as shown in Fig. 3.47. The equation for the voltage across the inductor L_1 is derived from Eq. (3.16) and written as follows:

$$V_{L1} = L_1 \cdot \frac{di_{L1}}{dt}_{(DT)}$$

By Eq. (3.32), $V_{L1} = V_{in}$, so the above equation can be written as follows:

$$L_1 \cdot \frac{di_{L1}}{dt}_{(DT)} = V_{in}$$

$$\frac{di_{L1}}{dt}_{(DT)} = \frac{V_{in}}{L_1} \tag{3.38}$$

Eq. (3.38) is the current change through the inductor L_1 when the switches are in position 1. Similarly during T_{OFF}.

$$V_{L1} = L_1 \cdot \frac{di_{L1}}{dt}_{(D'T)}$$

By Eq. (3.34), $V_{L1} = V_{in} - V_1$, so the above equation can be written as follows:

$$L_1 \cdot \frac{di_{L1}}{dt}_{(D'T)} = V_{in} - V_1$$

$$\frac{di_{L1}}{dt}_{(D'T)} = \frac{V_{in} - V_1}{L_1} \tag{3.39}$$

Eq. (3.39) is the current change through the inductor L_1 when the switches are in position 2.

To determine the relation for the inductor L_2, we need to calculate the rate of change of current through the inductor L_2 during T_{ON} and T_{OFF}, as shown in Fig. 3.48. The voltage across the inductor L_2 is given by the equation.

$$V_{L2} = L_2 \cdot \frac{di_{L2}}{dt}_{(DT)}$$

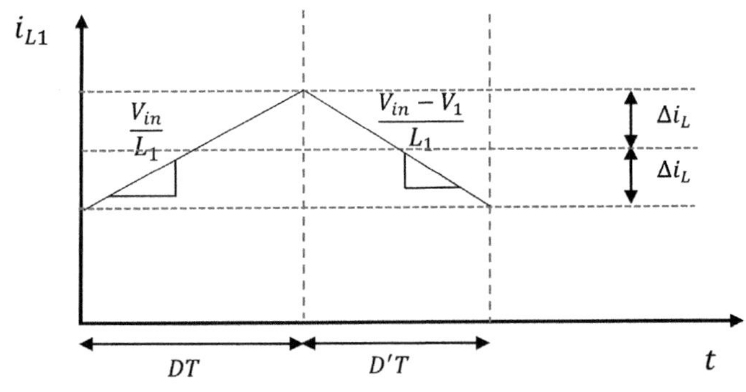

FIGURE 3.47 Inductor current (i_{L1}) in DC–DC Cuk converter.

$$\text{Change in } i_L = (\text{slope})(\text{length of sub} - \text{interval})$$

$$2\Delta i_{L1} = \frac{V_{in}}{L_1} \cdot DT$$

$$L_1 = \frac{V_{in}}{2\Delta i_{L1}} \cdot DT \tag{3.40}$$

By Eq. (3.33), $V_{L2} = -V_1 - V_2$, so the above equation can be written as following

$$L_2 \cdot \frac{di_{L2}}{dt}_{(DT)} = -V_1 - V_2$$

$$\frac{di_{L2}}{dt}_{(DT)} = \frac{-V_1 - V_2}{L_2} \tag{3.41}$$

Eq. (3.41) is the current change through the inductor L_2 when the switches are in position 1. Similarly during T_{OFF}.

$$V_{L2} = L_2 \cdot \frac{di_{L2}}{dt}_{(DT)}$$

By Eq. (3.35), $V_{L2} = -V_2$, so the above equation can be written as following

$$L_2 \cdot \frac{di_{L2}}{dt}_{(D'T)} = -V_2$$

$$\frac{di_{L2}}{dt}_{(D'T)} = \frac{-V_2}{L_2} \tag{3.42}$$

Eq. (3.42) is the current change through the inductor L_2 when the switches are in position 2.

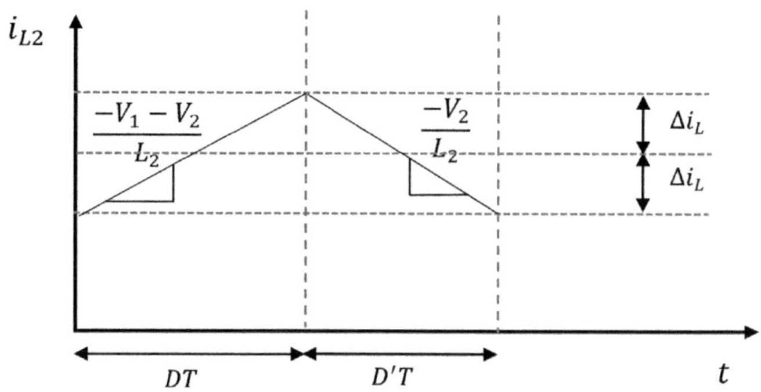

FIGURE 3.48 Inductor current (i_{L2}) in DC–DC Cuk converter.

$$\text{Change in } i_L = (\text{slope})(\text{length of sub} - \text{interval})$$

$$2\Delta i_{L2} = \frac{-V_1 - V_2}{L_2} \cdot DT$$

$$L_2 = \frac{-V_1 - V_2}{2\Delta i_{L2}} \cdot DT \tag{3.43}$$

3.5 Inverters (DC–AC inverters)

3.5.1 H-Bridge inverter

In daily life applications where we have a DC voltage source that needs to entertain the AC loads, an inverter is required to fulfill the requirement. This can be done by using the converters that can produce the voltage of both polarities [7]. In Section 3.4 we have seen that buck converters and boost converters produce only positive polarity voltages, as shown in Eqs. (3.15) and (3.22), respectively, whereas buck-boost and Cuk converters generate only negative polarity voltages, as shown in Eqs. (3.28) and (3.37). The bipolar or the AC voltage at the output can be obtained by connecting two buck converters differentially. The switching pattern must be such that if one converter is ON, the other must be OFF. Differentially connecting two buck converters and after simplification, Fig. 3.49 shows the circuit of an H-bridge inverter with SPDT switches. The SPDT switch can be realized by replacing a SPDT switch with two solid-state switches. The switch S_1 is replaced by Q_1 and Q_4, whereas S_2 is replaced by Q_2 and Q_3, as shown in Fig. 3.50. The central point or the junction of Q_1 and Q_4 is named as point A and is connected to the left side of the load. The mid of Q_2 and Q_3 is named as point B and is connected to the right side of the load.

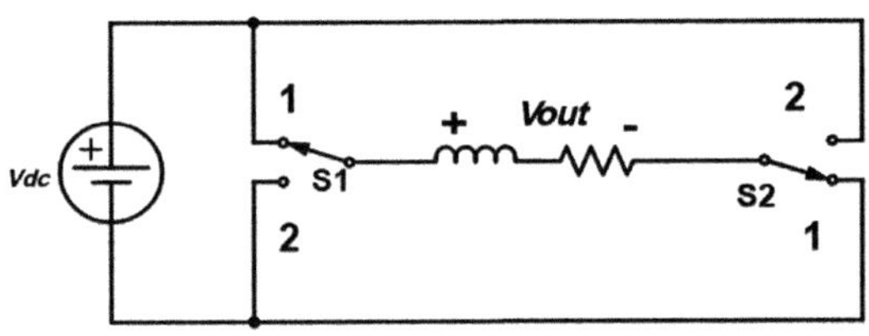

FIGURE 3.49 H-bridge with single-pole double-throw switches.

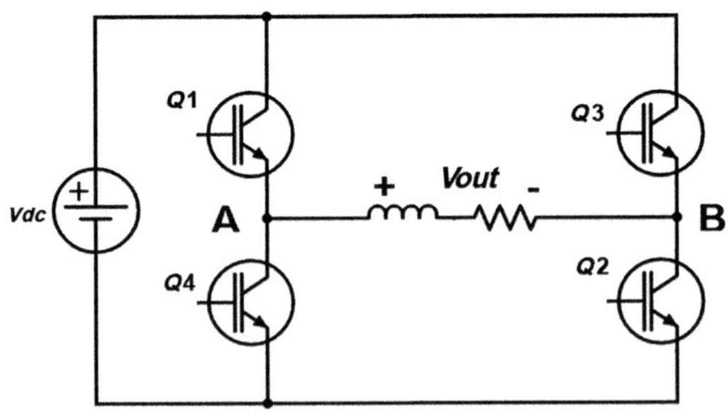

FIGURE 3.50 H-bridge inverter with switches realization.

For the proper operation of the inverter, if the upper switch from one leg is ON than the lower switch from the other leg must be OFF. When the switches Q_1 and Q_2 are ON and Q_3 and Q_4 are OFF, the current flows through the load from the positive end A to the negative end B. If the switches Q_3 and Q_4 are ON and Q_1 and Q_2 are OFF, the current flows through the load from the negative end B to the positive end A. This opposite flow of currents constitutes the AC signal.

3.5.2 Multilevel inverter

3.5.2.1 *Diode clamped multilevel inverters*

Multilevel inverters (MLI) can be of any level and the number of circuit elements needed to build up the diode clamped multilevel inverter (DC−MLI) can be found by using the following relations [8]. For an "n"-level inverter, the number of voltage sources (N_{dc}) can be found by Eq. (3.44), number of DC bus capacitors (N_C) by Eq. (3.45), number of clamping diodes (N_D) by Eq. (3.46), and number of switches $\left(N_Q\right)$ can be determined by using Eq. (3.47).

$$N_{dc} = (n - 1) \tag{3.44}$$

$$N_C = (n - 1) \tag{3.45}$$

$$N_D = (n - 1)(n - 2) \tag{3.46}$$

$$N_Q = 2(n - 1) \tag{3.47}$$

For a three-level inverter, the equations decided that two capacitors, two diodes, and four switches are needed in an arrangement shown in Fig. 3.51A. The capacitors C_1 and C_2 connected in series divide the input voltage V_{dc} into three levels $V_{dc}/2$, 0, and $-V_{dc}/2$. The switches Q_1, and Q_2 are activated for voltage level $V_{dc}/2$, Q_3 and Q_4 are activated for voltage level $-V_{dc}/2$, and Q_2 and Q_3 are activated for 0 voltage level, as shown in Fig. 3.51B. The switching scheme of a three-level DC−MLI is shown in Table 3.1.

Likewise, if we select the value of n as 5 for a five-level flying capacitor multilevel inverter (FC-MLI), we can use the equations to determine the numbers of capacitors, diodes, and switches for a five-level DC−MLI. For five-level DC−MLI we need four capacitors, 12 diodes, and eight switches. The voltage across each capacitor is $\frac{V_{dc}}{4}$, hence, the voltage levels for a five-level MLI are $\frac{V_{dc}}{2}$, $\frac{V_{dc}}{4}$, 0, $\left(\frac{-V_{dc}}{4}\right)$, and $\frac{-V_{dc}}{2}$. Each voltage level is associated with different switching states of the switches ($Q_1 - Q_8$) shown in Table 3.2. No. of diodes, no. of switches, and no. of capacitors required to build DC-MLI for different voltage levels are given in Table 3.3.

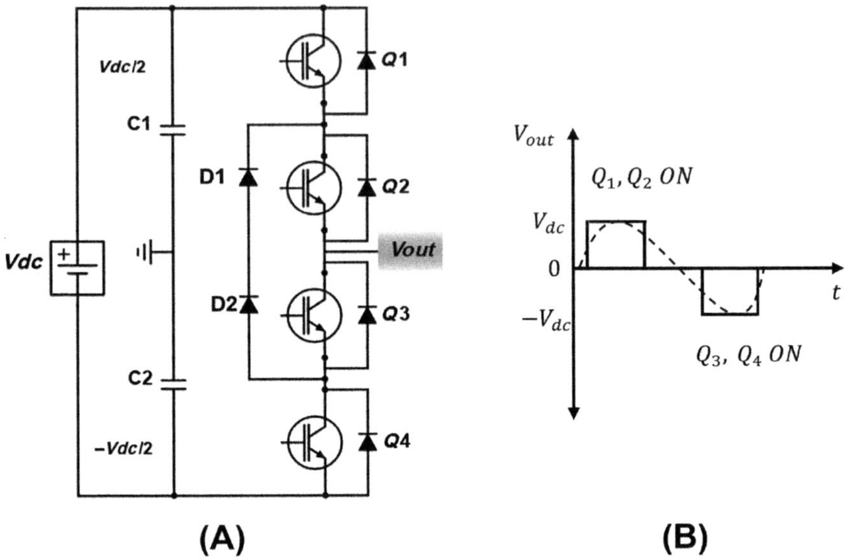

FIGURE 3.51 (A) Three-level diode clamped multilevel inverter (DC−MLI). (B) Voltage levels and switching states of three-level DC−MLI.

TABLE 3.1 Switching states and corresponding voltage levels of a three-level diode clamped multilevel inverter.

Voltages	V_{dc}	0	$-V_{dc}$
Switches			
Q_1	1	0	0
Q_2	1	1	0
Q_3	0	1	1
Q_4	0	0	1

For voltage level $\frac{V_{dc}}{2}$: switches (Q_1-Q_4) are kept ON, whereas the remaining ($Q_5 - Q_8$) are OFF. The source voltage V_{dc} is connected to the load through the capacitors C_1 and C_2, and switches Q_1, Q_2, Q_3, and Q_4.

For voltage level $\frac{V_{dc}}{4}$: switches ($Q_2 - Q_5$) are kept ON, whereas the remaining ($Q_1, Q_6, Q_7,$ and Q_8) are OFF. The source voltage V_{dc} is connected to the load through the capacitors C_2, diode D1 and switches, Q_2, Q_3, Q_4 and Q_5. As only C_2 is connected, the voltage level is $\frac{V_{dc}}{4}$.

For zero voltage level: switches ($Q_3 - Q_6$) are kept ON, whereas the remaining ($Q_1, Q_2, Q_7,$ and Q_8) are OFF. Therefore no capacitor is connected,

TABLE 3.2 Switching states and corresponding voltage levels of a five-level diode clamped multilevel inverter.

Voltages / Switches	$\frac{V_{dc}}{2}$	$\frac{V_{dc}}{4}$	**0**	$-\frac{V_{dc}}{4}$	$-\frac{V_{dc}}{2}$
Q_1	1	0	0	0	0
Q_2	1	1	0	0	0
Q_3	1	1	1	0	0
Q_4	1	1	1	1	0
Q_5	0	1	1	1	1
Q_6	0	0	1	1	1
Q_7	0	0	0	1	1
Q_8	0	0	0	0	1

TABLE 3.3 Circuit elements for different voltage levels of diode clamped multilevel inverter (DC–MLI).

Level *n*	Capacitors (N_C)	Diodes (N_D)	Switches (N_Q)
3	2	2	4
5	4	12	8
7	6	30	12
9	8	56	16
11	10	90	20
13	12	132	24

and both sides of the switches are directly connected to the ground maintaining a zero voltage level across the load.

For voltage level $-\frac{V_{dc}}{4}$: switches ($Q_4 - Q_7$) are kept ON, whereas the remaining ($Q_1, Q_2, Q_3,$ and Q_8) are OFF. Capacitor C_3 is connected to the load and ground through the switches $Q_4, Q_5, Q_6,$ and Q_7. The polarity of the voltage is negative as the current flows from the ground to the load.

For voltage level $-\frac{V_{dc}}{2}$: switches (Q_5–Q_8) are kept ON, whereas the remaining Q_1, Q_2, Q_3, and Q_4 are OFF. Capacitor C_3 and C_4 are connected to the load and ground through the switches Q_5, Q_6, Q_7 and Q_8. Again, the polarity of the voltage is negative as the current flows from the ground to

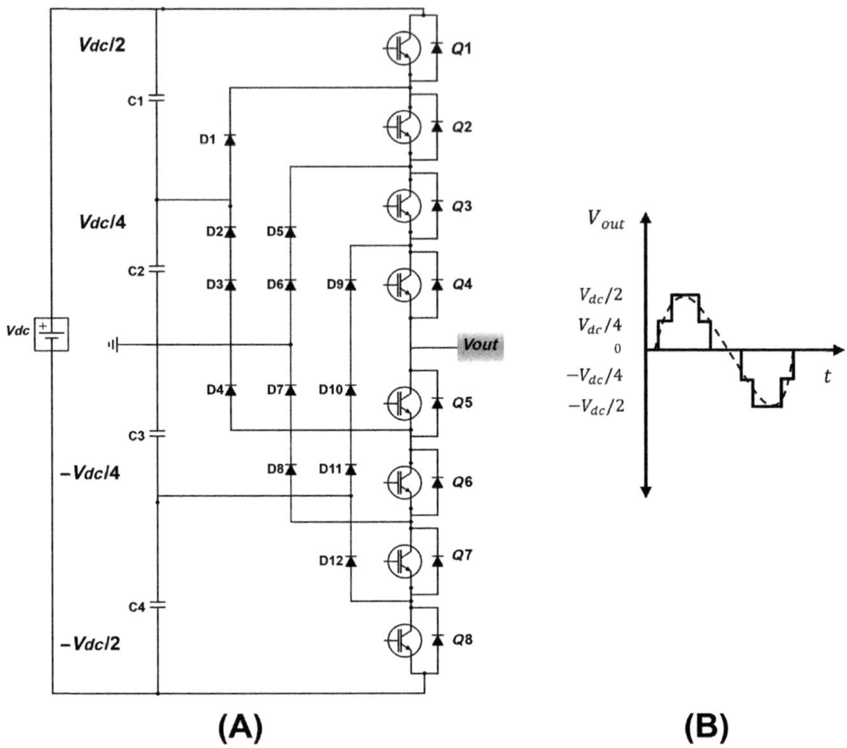

FIGURE 3.52 (A) Five-level diode clamped multilevel inverter (DC–MLI). (B) Voltage levels and switching states of five-level DC–MLI.

the load. The arrangement of circuit elements for a five-level DC–MLI is shown in Fig. 3.52A and the voltage levels and corresponding waveform are shown in Fig. 3.52B.

3.5.2.2 Cascaded H-Bridge multilevel inverters

Another topology of MLI with comparatively fewer components is the cascaded H-bridge multilevel inverter (HB-MLI). It is obtained by a series connection of H-bridge inverters with dedicated DC voltage sources. The DC sources for cascaded HB-MLI need necessarily to be isolated which makes it feasible for use with solar photovoltaic cells and fuel cells. The minimum and maximum voltage levels can be obtained for the given specific level of the inverter by using Eqs. (3.48) and (3.49) respectively. The number of switches and the number of input capacitors are determined by Eqs. (3.50) and (3.51), respectively.

$$V_{min} = (-V_{dc})\frac{n-1}{2} \tag{3.48}$$

$$V_{max} = (V_{dc})\frac{n-1}{2} \tag{3.49}$$

$$N_Q = 2(n-1) \tag{3.50}$$

$$N_C = \frac{n-1}{2} \tag{3.51}$$

The simple H-bridge inverter that is also called a three-level HB-MLI has already been discussed in detail. Fig. 3.53A shows the five-level cascaded H-bridge MLI that is a series connection of two H-bridge inverters. The five voltage levels $2V_{dc}, V_{dc}, 0, -V_{dc}, -2V_{dc}$ are shown in Fig. 3.53B. Eqs. (3.50) and (3.51) show that eight switches and two capacitors are required to construct the five-level CHB-MLI. The circuit diagram of the five-level CHB-MLI is shown in Fig. 3.53A. The switching states corresponding to each voltage level are shown in Table 3.4.

For voltage level $2V_{dc}$ at the output: switches $Q_1, Q_2, Q_5,$ and Q_6 must be ON, while the switches $Q_3, Q_4, Q_7,$ and Q_8 must essentially be OFF. For this voltage level, current adopts the path $V_{dc1}, Q_1,$ load, $Q_6, V_{dc2}, Q_5, Q_2,$ and V_{dc1}. As both voltage sources are in the path of the current, we get their sum at the output.

For voltage level V_{dc} at the output: switches $Q_1, Q_2, Q_6,$ and Q_8 must be ON, while the switches $Q_3, Q_4, Q_5,$ and Q_7 must essentially be OFF. For this

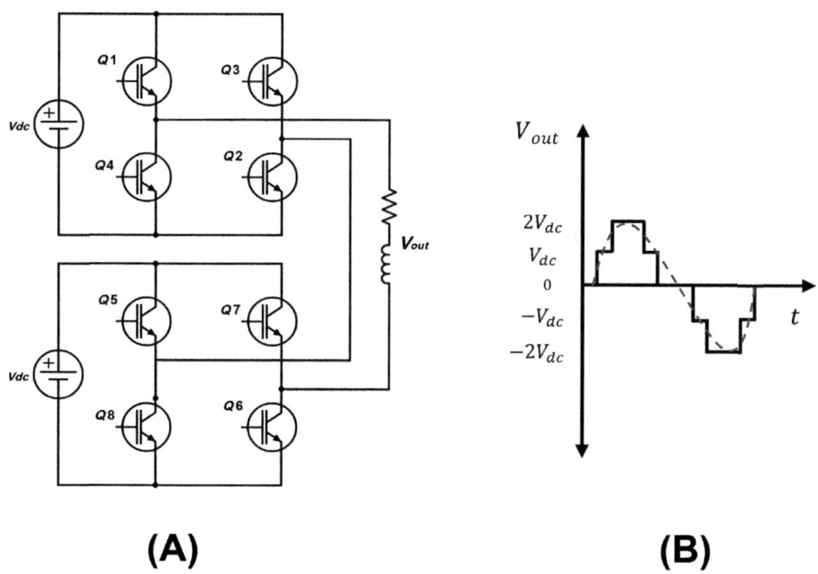

FIGURE 3.53 (A) Five-level cascaded H-bridge multilevel inverter (HB-MLI). (B) Voltage levels and switching states of five-level cascaded HB-MLI.

TABLE 3.4 Switching states and corresponding voltage levels of a five-level cascaded H-bridge multilevel inverter (HB-MLI).

Voltages Switches	$2V_{dc}$	V_{dc}	0	$-V_{dc}$	$-2V_{dc}$
Q_1	1	1	0	0	0
Q_2	1	1	1	0	0
Q_3	0	0	0	1	1
Q_4	0	0	1	1	1
Q_5	1	0	0	0	0
Q_6	1	1	1	1	0
Q_7	0	0	0	0	1
Q_8	0	1	1	1	1

voltage level, current adopts the path $V_{dc1}, Q_1,$ load, $Q_6, Q_8, Q_2,$ and V_{dc1}. As the second voltage source is bypassed, we get only V_{dc} at the output.

For zero voltage level at the output: switches $Q_2, Q_4, Q_6,$ and Q_8 must be ON, while the switches $Q_1, Q_3, Q_5,$ and Q_7 must essentially be OFF. In the followed current path, not a single source is connected to the load.

For voltage level $-V_{dc}$ at the output; switches $Q_3, Q_4, Q_6,$ and Q_8 must be ON, while the switches $Q_1, Q_2, Q_5,$ and Q_7 must essentially be OFF. For this voltage level, current adopts the path $V_{dc1}, Q_3, Q_8, Q_6,$ load $Q_4,$ and V_{dc1}. As the second voltage source is bypassed, we get only $-V_{dc}$ at the output. The negative sign indicates that the current in this path flows in the reverse direction from the load.

For voltage level $-2V_{dc}$ at the output: switches $Q_3, Q_4, Q_7,$ and Q_8 must be ON, while the switches $Q_1, Q_2, Q_5,$ and Q_6 must essentially be OFF. For this voltage level, current adopts the path $V_{dc1}, Q_3, Q_8, V_{dc2}, Q_7$ load, and Q_4 As both voltage sources come in the path of the current, so we get their sum at the output. The negative sign indicates that the current in this path flows in the reverse direction from the load.

3.5.2.3 Flying capacitor multilevel inverters

In a comparatively new topology instead of diodes capacitors are used to clamp the voltage. These MLIs can be of any level and the number of circuit elements needed to build up the FC-MLI can be determined by using the following relations. For an n-level inverter, the number of voltage sources (N_{dc}) can be found by Eq. (3.52), the number of DC bus capacitors (N_C) by Eq. (3.53), the number of balancing capacitors (N_{BC}) by Eq. (3.54), and the number of switches (N_Q) can be determined by using Eq. (3.55). The voltage across each capacitor is the same for a given level as in DC–MLI, however, the switching combination may be different from DC–MLI.

$$N_{dc} = (n - 1) \tag{3.52}$$

$$N_C = (n - 1) \tag{3.53}$$

$$N_{BC} = (n - 1)(n - 2)/2 \tag{3.54}$$

$$N_Q = 2(n - 1) \tag{3.55}$$

For a three-level FC-MLI, the above equations Eqs. (3.52) to (3.55) determined that two DC bus capacitors, one balancing capacitors, and four switches are required in an arrangement shown in Fig. 3.54A. The capacitors C_1 and C_2 connected in series divide the input voltage V_{dc} into three levels $V_{dc}/2$, 0, and $-V_{dc}/2$. As the input voltage V_{dc} is connected to the series combination of C1 and C2, it is developed equally across each capacitor. The switches Q_1 and Q_2 are activated for voltage level $V_{dc}/2$, Q_3 and Q_4 are activated for voltage level $-V_{dc}/2$, and Q_1 and Q_3 or Q_2 and Q_4 are activated for 0 voltage level. The voltage levels and the corresponding switching states are shown in Fig. 3.54B. The switching scheme of a three-level FC-MLI is shown in Table 3.5.

Similarly, if we select the value of n as 5, we can use the equations to determine the numbers of DC bus capacitors, balancing capacitors, and switches for a five-level DC–MLI. For a five-level FC-MLI we need four DC bus capacitors, six balancing capacitors, and eight switches. The arrangement of circuit elements of a five-level FC-MLI is shown in Fig. 3.55A. The

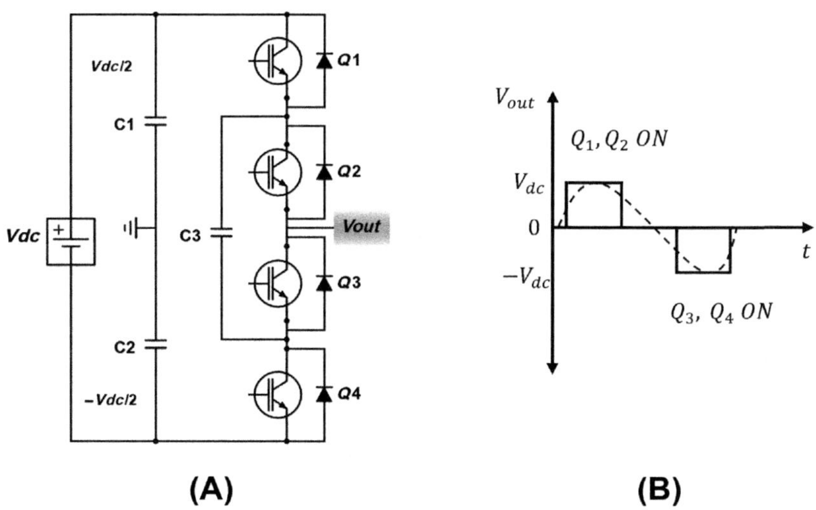

FIGURE 3.54 (A) Three-level flying capacitor multilevel inverter (FC-MLI). (B) Voltage levels and switching states of three-level FC-MLI.

TABLE 3.5 Switching states and corresponding voltage levels of a three-level flying capacitor multilevel inverter.

Voltages / Switches	V_{dc}	0	$-V_{dc}$
Q_1	1	1	0
Q_2	1	0	0
Q_3	0	1	1
Q_4	0	0	1

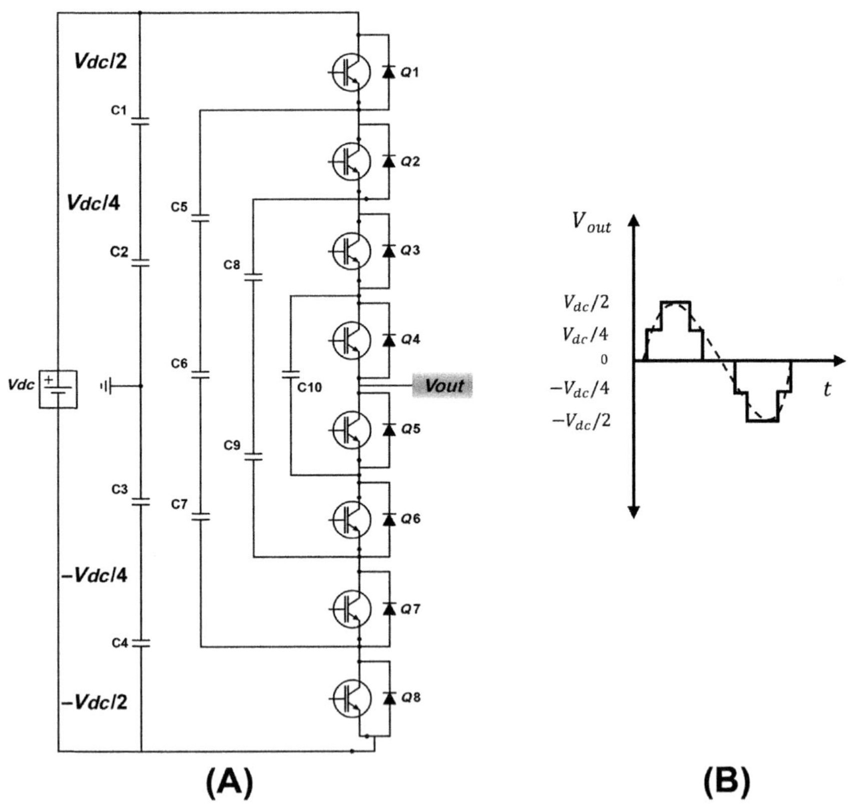

FIGURE 3.55 (A) Five-level flying capacitor multilevel inverter (FC-MLI). (B) Voltage levels and switching states of five-level FC-MLI.

TABLE 3.6 Switching states and corresponding voltage levels of a five-level flying capacitor multilevel inverter.

Voltages	$\frac{V_{dc}}{2}$	$\frac{V_{dc}}{4}$	0	$-\frac{V_{dc}}{4}$	$-\frac{V_{dc}}{2}$
Switches					
Q_1	1	1	1	1	0
Q_2	1	1	1	0	0
Q_3	1	1	0	0	0
Q_4	1	0	0	0	0
Q_5	0	1	1	1	1
Q_6	0	0	1	1	1
Q_7	0	0	0	1	1
Q_8	0	0	0	0	1

voltage across each capacitor is $\frac{V_{dc}}{4}$, hence the voltage levels for a five-level FC-MLI are $\frac{V_{dc}}{2}$, $\frac{V_{dc}}{4}$, 0, $\left(\frac{-V_{dc}}{4}\right)$, and $\frac{-V_{dc}}{2}$. Each voltage level is associated with different switching states of the switches ($Q_1 - Q_8$) shown in Fig. 3.55 (b). For voltage level $\frac{V_{dc}}{2}$: switches ($Q_1 - Q_4$) are kept ON, whereas the remaining ($Q_5 - Q_8$) are OFF. For voltage level $\frac{V_{dc}}{4}$: switches $Q_1, Q_2, Q_3,$ and Q_5 are ON, whereas the remaining ($Q_4, Q_6, Q_7,$ and Q_8) are OFF. For zero voltage level: switches $Q_1, Q_2, Q_5,$ and Q_6 are ON, whereas the remaining $Q_3, Q_4, Q_7,$ and Q_8 are OFF. For voltage level $-\frac{V_{dc}}{4}$: switches $Q_1, Q_5, Q_6,$ and Q_7 are ON, whereas the remaining $Q_2, 3, Q_4,$ and Q_8 are OFF. For voltage level $-\frac{V_{dc}}{2}$: switches Q_5, Q_6, Q_7, and Q_8 are ON, whereas the remaining Q_1, Q_2, Q_3, and Q_4 are OFF. The switching states for different voltage levels in a five-level FC-MLI are summarized in Table 3.6.

3.6 Cycloconverters (AC–AC converters)

Transformers are used to step up or step down the voltage level of an AC voltage without changing the frequency. The frequency of the output signal is the same as of the input AC signal. In various applications like rolling mills, water pumps, washing machine, grinding mills, and other industrial applications, we need an AC signal of different frequency. For this purpose we use a cycloconverter that not only changes the voltage level but also the frequency of the input AC signal without using an intermediate DC link [9]. Fig. 3.56 shows the circuit diagram of a single-phase to single-phase thyristor-based cycloconverter. It consists of back to back thyristor-based full bridge rectifier and inverted full bridge rectifier. The bridge located at the left side of the load is called the positive rectifier and the bridge located at

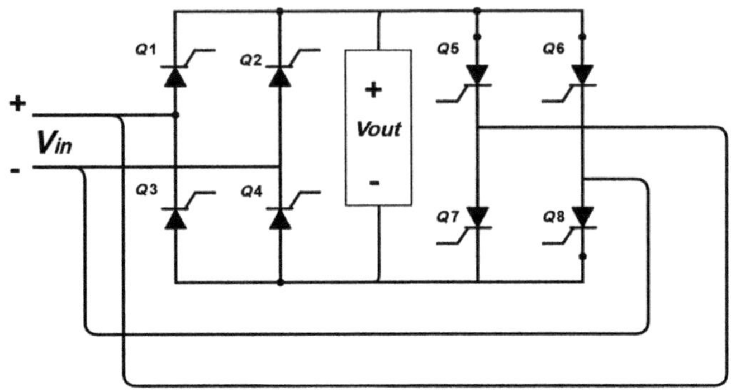

FIGURE 3.56 AC–AC cycloconverter.

the right side of the load is called the negative rectifier as shown in Fig. 3.56.

If we need to have an AC signal of half of the input frequency, we will have a specific switching scheme. During the first cycle of the input signal, a positive converter must be in operation by turning the Q_1 and Q_4 ON in first positive cycle. The current follows the path + Vin, Q_1, + load, Q_4, and −Vin. During the first negative half cycle, Q_2 and Q_3 must be conducting and the current follows the path −Vin, Q_2, + load, Q_3, and + Vin. In this way one complete cycle of the input signal appears as the positive cycle at the output. During the second cycle of the input signal, the negative converter must be in operation by turning the Q_6 and Q_7 ON in positive cycle. The current follows the path + Vin, Q_7, −load, Q_6, and −Vin. During the negative half cycle, Q5 and Q_8 must be conducting and the current follows the path −Vin, Q_8, −load, Q_3, and + Vin. In this way one complete cycle of the input signal appears as the negative cycle at the output. From Fig. 3.57 it can be seen that the input AC signal with frequency two cycles/second (2 Hz) has been converted into the one cycle/second AC signal shown by the dotted line. This is called a step down cycloconverter.

A step up cycloconverter transforms the input AC signal into a high frequency AC output. If we need an output AC signal that has a fourth time high frequency of the input AC signal, each half cycle of the input signal is divided into four equal intervals. During the positive half cycle, switching pairs Q_1 and Q_4 from the positive converter and Q_6 and Q_7 from the negative converter are alternately ON and OFF, as shown in Fig. 3.58. Similarly, during the negative half cycle, switching pairs Q_2 and Q_3 from the positive converter and Q_5 and Q_8 from negative converter are alternatively ON and OFF, as shown in Fig. 3.58. Output waveforms in Fig. 3.58 show that the frequency of the output is four times the frequency of the input AC signal. The cycles of the output are from t_1 to t_3, t_3 to t_5, t_5 to t_7, and t_7 to t_9, making a 4 Hz frequency signal.

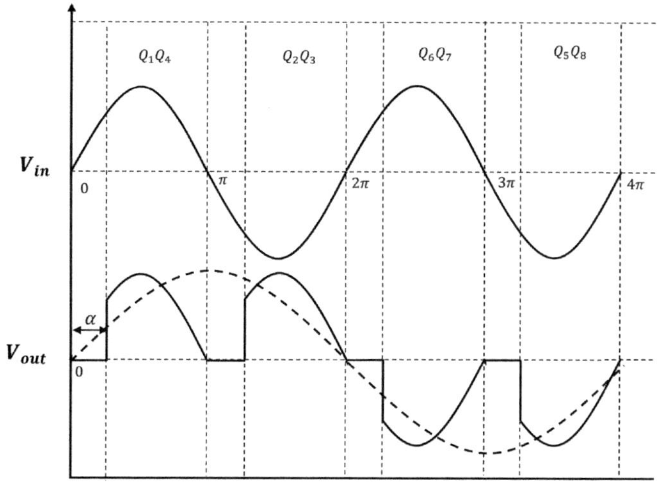

FIGURE 3.57 Output waveform of a step down cycloconverter.

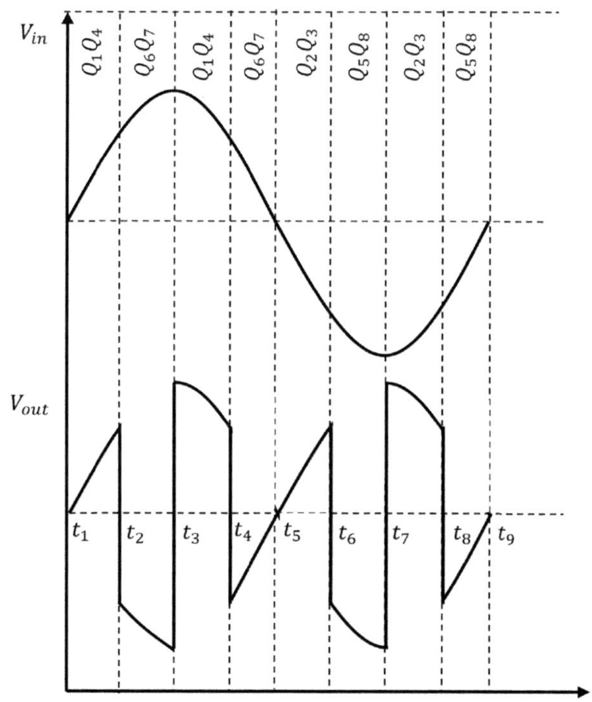

FIGURE 3.58 Output waveform of a step up cycloconverter.

Problems

Problems 1–10 contain four answer options A, B, C, and D. Choose the correct answer.

1. Which of the following can be turned ON by the gate terminal but cannot be turned OFF by the gate terminal?
   A. SCR
   B. SCS
   C. GTO
   D. TRIAC
2. Which of the following can be turned ON and OFF by the gate terminal?
   A. SCR
   B. SCS
   C. GTO
   D. TRIAC
3. Which of the following contains two gate terminals?
   A. SCR
   B. SCS
   C. GTO
   D. TRIAC
4. Which of the following is a gateless device?
   A. SCR
   B. SCS
   C. DIAC
   D. TRIAC
5. Which of the following is a bidirectional device?
   A. SCR
   B. SCS
   C. DIAC
   D. TRIAC
6. A solid-state switch is a
   A. Current controlled device
   B. Voltage controlled device
   C. Both A and B
   D. None of the above
7. In DC choppers, a SPDT switch is realized by
   A. An active switch
   B. A passive switch
   C. An active and a passive switch
   D. A diode
8. The number of switches in a five-level DC–MLI is
   A. 10
   B. 90

C. 20
D. 25

9. The output of a solar panel is to be fed to the grid, which device will be used?
A. Rectifier
B. Inverter
C. Chopper
D. SCR

10. If the input voltage of a Cuk converter is 6 V. At 0.7 duty cycle, what would be the output voltage?
A. 10 V
B. 14 V
C. −14 V
D. −10 V

11. What is the gate triggering method of SCR?

12. Differentiate between latching current and holding current.

13. What are the advantages of a freewheeling diode in a phase controlled half-wave rectifier with inductive load?

14. What are the step up and step down choppers? Differentiate them with an input and output relation.

15. A step up chopper is increasing the voltage level from 12 V to 24 V with a duty ratio 0.5 and a switching time period of 2.5 ms. Calculate the average inductor voltage.

16. What is the difference between a transformer and a cycloconverter?

References

[1] J. Pollefliet, 4 - Thyristors, Power Electronics: Switches and Converters. in: Pollefliet (Ed.), Academic Press, 2018, pp. 4.1−4.50.

[2] M.H. Rashid, 7 - Gate Turn-off Thyristors, in: Rashid (Ed.), Power Electronics Handbook, Butterworth-Heinemann, Boston, 2011, pp. 117−123.

[3] I. Grout, 8 - Interfacing Digital Logic to the Real World: A/D Conversion, D/A Conversion, and Power Electronics, in: Digital system design with Fpga, Cpld. Grout, Newnes, Burlington, 2008, pp. 537−614.

[4] J. Pollefliet, 8 - Controlled Rectifiers, Power Electronics: Switches and Converters. in: Pollefliet (Ed.), Academic Press, 2018, pp. 8.1−8.58.

[5] J. Pollefliet, 12 - Choppers, Power Electronics: Switches and Converters. in: Pollefliet (Ed.), Academic Press, 2018, pp. 12.1−12.14.

[6] H. Shahid, M. Kamran, Z. Mehmood, M.Y. Saleem, M. Mudassar, K. Haider, Implementation of the novel temperature controller and incremental conductance MPPT algorithm for indoor photovoltaic system, Sol. Energy (2018). Available from: https://doi.org/10.1016/j.solener.2018.02.018.

[7] J. Pollefliet, 14 - Inverters, Power Electronics: Switches and Converters, in: Pollefliet (Ed.), Academic Press, 2018, pp. 14.1−14.42.

[8] A. Sinha, K. Chandra Jana, M. Kumar Das, An inclusive review on different multi-level inverter topologies, their modulation and control strategies for a grid connected photovoltaic system, Sol. Energy 170 (2018) 633–657. Available from: https://doi.org/10.1016/j.solener.2018.06.001.

[9] J. Pollefliet, 10 - Cycloconverters, Power Electronics: Switches and Converters. in: Pollefliet (Ed.), Academic Press, 2018, pp. 10.1–10.6.

Chapter 4

Solar energy

Muhammad Kamran
Department of Electrical Engineering and Technology, Riphah International University, Pakistan

4.1 Introduction

Since the dawn of mankind, energy has been used in one or the other form. In the ancient days, the energy of the sun was used without any technology or apparatus. Later, to fulfill energy needs, conventional fossil fuels were used for cooking, heating, and electricity needs. Now with time, conventional energy sources are depleting, and it is time to shift ourselves to renewable energy. Almost all renewable energy sources are basically because of solar energy. The sun is a giant source of energy. Wind differential is created because of the difference in temperature of the air at two different locations. The difference in temperature causes a difference in the density of the air. The air with lower density rises. To fill this gap, wind from other regions moves to this region causing the flow of wind that is used to rotate the wind turbine.

Hydropower is another renewable energy source that exists because of solar energy. The water from the oceans, rivers, canals, and other places evaporates after receiving energy from solar radiation and migrates to other regions generating clouds. In this way, the water in the form of precipitation reaches the glaciers and comes back to the rivers where large dams or micro hydropower plants can be constructed. Biomass, a renewable energy source completes its life cycle because of the sunlight. The crops, fruits, and vegetables use the water from the soil and prepare their food in the presence of sunlight by the process of photosynthesis. This biomass is used as an energy source and it can replenish itself by taking water from the ground and the light from the sun.

Solar energy is directly utilized in two ways: solar thermal energy and solar photovoltaics (PVs). In solar thermal energy solar radiation is used to raise the temperature of the working fluid that can be further used for space heating, process heating, and electricity generation. In solar PVs, solar radiation is used to knock the electrons from the valence band of the semiconductor materials to the conduction band. This flow of electrons in the conduction band generates DC electricity. Another use of solar energy is the

Renewable Energy Conversion Systems. DOI: https://doi.org/10.1016/B978-0-12-823538-6.00008-7

indoor lighting system (ILS). In ILS, solar radiation is concentrated on the inlet point of the optical fibers using Fresnel lenses. These optical fibers guide the sunlight indoor fulfilling the lighting requirements.

4.2 Solar thermal

Solar thermal energy is technology that uses solar energy for electricity generation and space and water heating. Solar thermal technologies include solar parabolic trough, solar towers, solar dish, solar water heater, solar dryer, and solar cooker [1].

4.2.1 Solar parabolic trough

Among all the Concentrated Solar Power (CSP) technologies, the solar parabolic trough is considered to be the most mature technology. The technology consists of parabolic troughs, reflectors, and solar field piping. Parabolic troughs act as a solar reflector and all the light reflected by the parabolic troughs is focused on the absorber tubes located at the focal line of the reflectors. Absorber tubes contain the working fluid that moves from the inlet of the tubes to the outlet and collects the energy from the sun's radiation. As the solar parabolic trough uses only the direct normal irradiance (DNI), the use of a solar tracker is inevitable [2]. The temperature of the working fluid reported in the literature is in the range of 500°C–550°C. Fig. 4.1 shows the arrangement of a solar parabolic trough.

Because of the high temperature, water is used as a working fluid to generate steam. If the water is used as a working fluid, the steam generated can be used to run the integrated steam turbine and the generator working on the Rankine cycle.

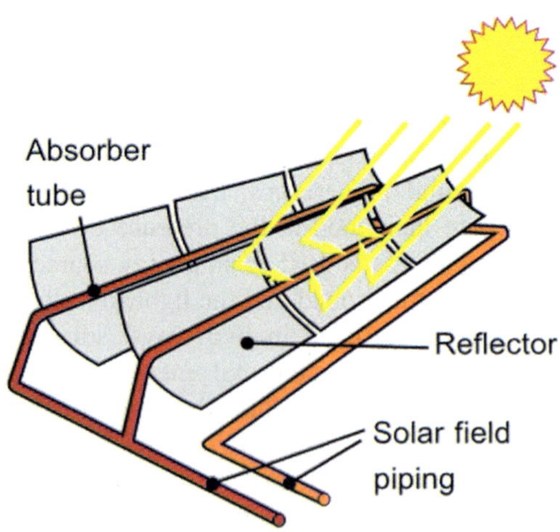

FIGURE 4.1 Solar parabolic trough.

The Rankine cycle integrated with the solar tower is explained in the chapter. If air is used as a working fluid, hot gases are produced by the solar receiver, and these gases are used to run the gas turbine working on the Brayton cycle.

4.2.2 Solar tower

The solar tower is a solar thermal technology consisting of a large solar energy collector mounted on the solar tower, multiple solar reflectors known as heliostats, thermal storage, and a generating unit. The heliostats are mounted on the dual-axis solar trackers that track the sun on the azimuthal angle and the altitude angle in a way that the solar radiation is reflected by them and focused on the solar energy receiver. The dual-axis solar trackers and the altitude and azimuthal angles are explained below. The performance characteristics and the efficiency of the solar tower power plant depend upon the working fluid. The most used working fluids are water, air, and molten salt. A solar tower is shown in Fig. 4.2.

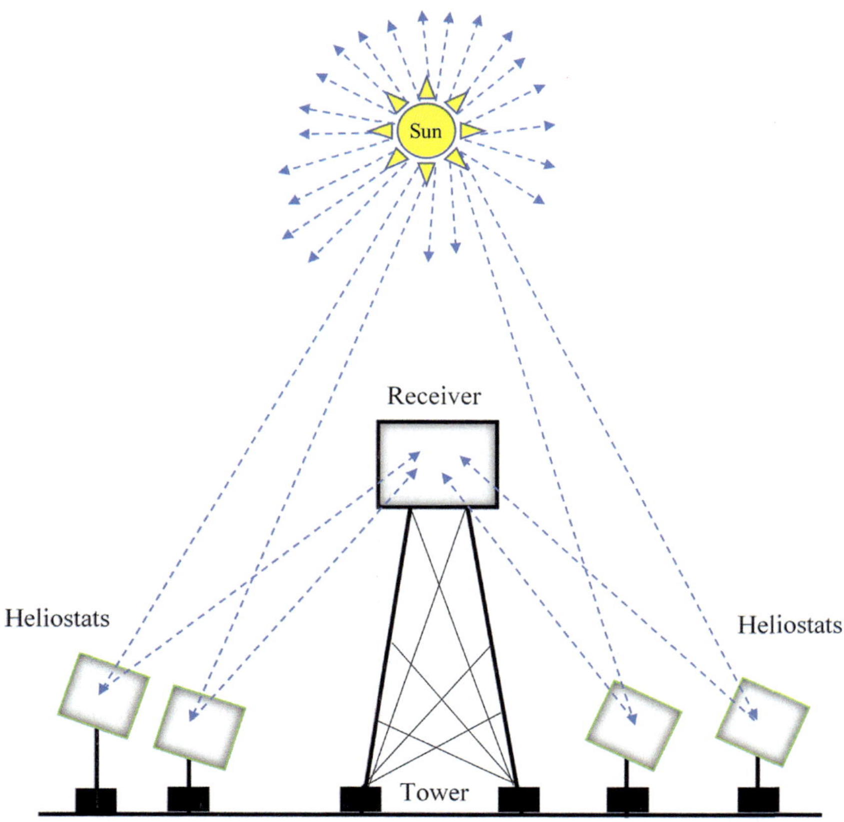

FIGURE 4.2 Solar tower.

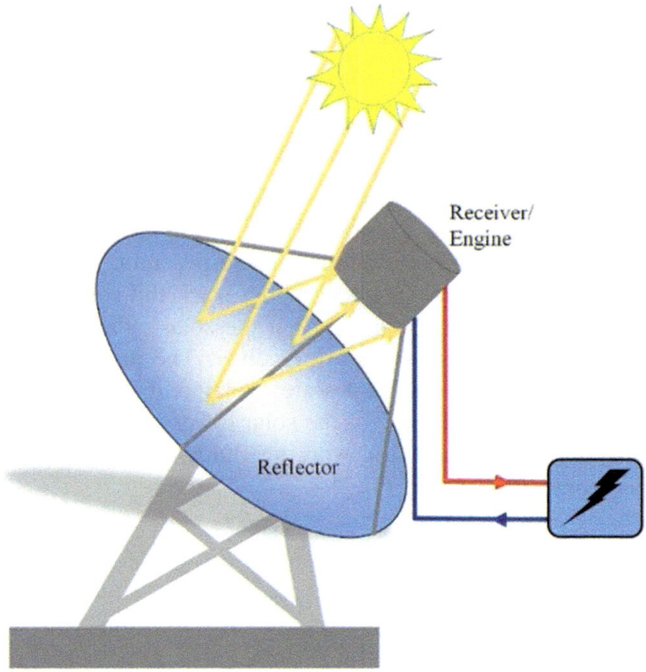

FIGURE 4.3 Solar parabolic dish.

If the water is used as a working fluid, the steam is generated by the receiver that can be used to run the integrated steam turbine and the generator working on the Rankine cycle. The Rankine cycle integrated with the solar tower is explained in the chapter. If air is used as a working fluid, hot gases are produced by the solar receiver, and these gases are used to run the gas turbine working on the Brayton cycle. The Brayton cycle is explained in the chapter.

4.2.3 Solar parabolic dish

The solar parabolic dish consists of a solar reflector dish, solar receiver, and the mounting structure. The dish is mounted on the dual-axis solar tracker that keeps it oriented always toward the sun. The solar radiation after being reflected by the dish are concentrated on the central receiver positioned at the focal point of the solar parabolic dish. The solar parabolic dish finds its major application in electricity generation using the Stirling engine mounted at the central receiver. Stirling engines have reported 30% solar to electricity efficiency, which is higher than other concentrated solar technologies. A solar parabolic dish is shown in Fig. 4.3.

4.2.4 Solar cooker

Solar cooking is a concept not limited to cooking food by the solar concentrated radiations but it also involves sterilization and pasteurization. It has replaced the direct incineration of the biomass materials that cause illness, pain, eye problems, and deforestation. Solar cookers also work on the principle of parabolic troughs that focus the sunlight at a central receiver where the cooking pot is placed to cook the food. Based on their construction and applications, solar cookers are categorized as solar box cooker, solar panel cooker, and solar parabolic cookers.

A solar box cooker consists of a reflective surface and a transparent glass window to allow the solar radiation to pass through into the insulated box. To maximize solar radiation absorption, the inner part of the box is painted black.

Solar panel cookers are the most simple, cheap, portable, and affordable solar cookers. They consist of simply a reflective element and the cooking vessel placed inside a plastic bag [3].

A solar parabolic cooker works on the principle of the solar parabolic trough. The solar radiations are concentrated by the reflector troughs and are directed to a point where the cooking vessel is placed. Because of the concentration of solar radiation, it may have the problem of burning the food if left unattended.

4.2.5 Solar water heater

The simplest application of solar thermal energy is the solar water heater that is used to heat the water for washing, heating, cleaning, and cooking. Solar water heaters are categorized as active and passive solar heaters. In passive solar heaters, collectors take the heat from the solar radiation and transfer it to the water in the water tank mounted on the top of the collectors. The hot water of the order of 60°C moves by the natural circulation process and is best suited for domestic applications. In the natural circulation process, cold water enters the water tank from the bottom takes the heat from the heat exchanger, and gets warmed. The density of the warmed water is lower so it moves upward to the hot water outlet pipe, as shown in Fig. 4.4.

In an active solar water heater, instead of the natural circulation, hot water is moved by the pump. If the water temperature of the water in the upper part of the tank is higher than the bottom part, the reverse flow of the water is prohibited by a check valve.

4.2.6 Solar dryer

Solar dryers are used to eliminate the moisture content from crops, vegetables, and fruits. The solar dryer consists of a box made up of easily available and cheap material like cement, galvanized iron, brick, and plywood. The top surface of the dryer is covered by transparent single and double-layered sheets. The inside surface is colored black to absorb the incoming solar radiation. Since the box is insulated, the inside temperature of the box is raised.

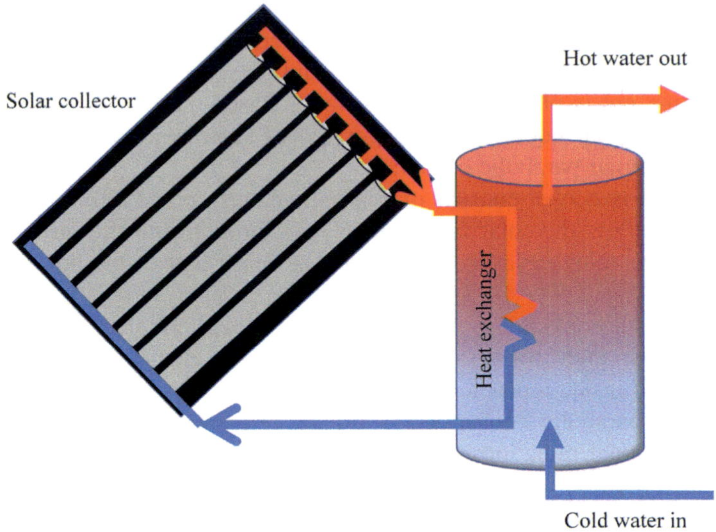

FIGURE 4.4 Solar water heater.

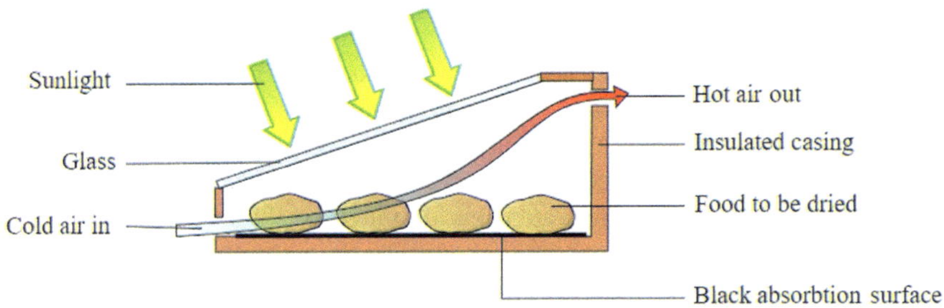

FIGURE 4.5 Solar dryer.

The air is ventilated through the small holes at the top of the box. As the inside air gets warm, it rises by the natural circulation process and removes the moisture from the fruits, vegetables, and the crops placed in trays inside the box. To fill the vacuum, fresh air comes in by a forced draught process and the process continues, as shown in Fig. 4.5.

4.3 Solar photovoltaic

Apart from the thermal utilization of the solar energy, it can also be used in semiconducting materials. It is well-known that the sun's energy consists of photons of light—packets of energy—corresponding to different wavelengths. This light could be absorbed, reflected back, or pass through the PV material. If it is absorbed by the material, it transfers its energy to electrons that leave the valence band and move in the external circuit to build a

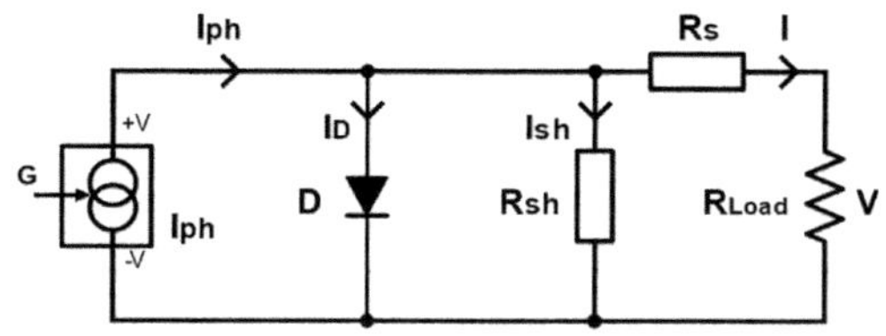

FIGURE 4.6 Single diode model of a solar cell.

voltage across the P-N junction. This effect is called the PV: development of voltage by light.

4.3.1 Modeling of PV cell

All the solar cells can be modeled by a current source, diode, and two different resistances representing the different parameters of the solar cell. When the solar radiation falls on the PV cell, electrons gain the sun's energy and are knocked off the valence band to the conduction band leaving behind the holes, so both move in the opposite direction, which can be represented by a current source. The movement of electrons generated by the solar irradiance is represented by a current source. Here, a single diode model of the PV solar cell, shown in Fig. 4.6, and equivalent mathematical equations are used to acquire the output curves of the PV cell.

4.3.1.1 Photocurrent

Photocurrent, also known as the short-circuit current, is the current generated by the sun's irradiance falling on the PV cell. The valence electrons absorb the photons having energy equivalent to their bandgap energy and are knocked out of the valence band and move in the conduction band. This movement of electrons in the conduction band is represented by a current source and is termed as I_{ph}.

4.3.1.2 Forward-biased diode

If the PV cell is not connected to the external load then there exists an open-circuit voltage that allows a full current to flow effectively through a P-N junction acting as a forward-biased diode. But when a load is connected, though the voltage is dropped still there is a voltage that causes the flow of current through the forward-biased diode. This action of the solar cell is modeled by a diode in parallel to the current source I_{ph} as shown in Fig. 4.6.

4.3.1.3 Series resistance

During the current flow from the generating source to the load, there are various types of resistance in its path, like the resistance between the top and rear metal contacts, silicon, and the metal contacts, and the resistance of the emitter to the base path. The sum of all those resistances faced by the current as it passes through the PV cell to the external metal contacts through bulk material and to the load is termed as series resistance, as it appears in series to the load, as shown in Fig. 4.6.

4.3.1.4 Shunt resistance

The energy in the photons of the solar irradiance is of a different level. If the energy of the photons matches the bandgap energy of the solar cell material, it knocks them off the holes from the valence band to the conduction band, generating a photocurrent. If the energy of the photons is barely equal to the bandgap energy of the cell material, it knocks the electron off from the valence band. But, before they cross the valence band to reach the conduction band, their energy is lost and they come back to the valence band. Electrons–holes recombination occurs causing the reduction in the originally generated current I_{ph} and this effect is termed as the shunt resistance in the equivalent diagram of the solar cell shown in Fig. 4.6. The bandgap energy of different semiconductor materials is given in Table 4.1.

4.3.1.5 Open-circuit voltage

Open-circuit voltage is the voltage across the solar cell when there is no current, meaning that the load is not connected. Any change in the open-circuit voltage indicates a variation in the solar irradiance. A relation shown in Eq. (4.1) for the open-circuit voltage is determined by applying the Kirchhoff current law at the single diode model of the solar cell.

$$I = I_{Ph} - I_D - I_{sh} \tag{4.1}$$

where I_{Ph} is photonic current, I_D is current through parallel diode D, I_{sh} is current through the shunt resistance R_{sh}, and I is the current through the

TABLE 4.1 Bandgap energy of different semiconductor materials.

Sr. no.	Semiconductor	E_{gap}(eV)
1	*Si*	1.12
2	*Ge*	0.67
3	*CdTe*	1.44
4	*CdZnTe*	1.6
5	*HgI$_2$*	2.13
6	*GaAs*	1.42

series resistance R_s. The value of the diode current is determined by the Shockley diode equation given by Eq. (4.2).

$$I_D = I_0\left[\left(e^{\frac{V_{sh}.q}{n.k.T}}\right) - 1\right] \tag{4.2}$$

where I_0 is saturation current, V_{sh} is the voltage across the shunt resistance, q is charge on an electron $\left(1.602\times10^{-19}\mathrm{C}\right)$, T is cell temperature (K), n is ideality factor of solar material, and$k$ is Boltzmann's constant $(1.381\times10^{-23}J/K)$. Current through the shunt resistor is calculated by Eq. (4.3).

$$I_{sh} = \frac{V_{sh}}{R_{sh}} = \frac{V + IR_s}{R_{sh}} \tag{4.3}$$

Putting the values of I_D and I_{sh} in Eq. (4.1), we get Eq. (4.4).

$$I = I_{Ph} - I_0\left[\left(e^{\frac{V_{sh}}{n.V_T}}\right) - 1\right] - \frac{V + IR_s}{R_{sh}} \tag{4.4}$$

At no load when the circuit is open, there is no current through the load, that is, $I = 0$. Thus there is no voltage drop at the series resistor (R_s) and the voltage at the output is the same as across the shunt resistor (R_{sh}) and we can denote it as V_{oc}.

$$0 = I_{Ph} - I_0\left[\left(e^{\frac{V_{OC}}{n.V_T}}\right) - 1\right] - \frac{V_{oc}}{R_{sh}} \tag{4.5}$$

The last term (V_{oc}/R_{sh})can be neglected as the R_{sh} is much larger than the V_{oc}. On rearranging Eq. (4.5) we have a relation for the open-circuit voltage shown in Eq. (4.6).

$$\frac{I_{Ph}}{I_0} = \left(e^{\frac{V_{OC}}{n.V_T}}\right) - 1$$

$$V_{oc} = ln\left(\frac{I_{Ph}}{I_0} + 1\right)\left(\frac{n.k.T}{q}\right) \tag{4.6}$$

Eq. (4.6) shows that the open-circuit voltage largely depends upon the photonic current generated by solar irradiance, the ideality factor of the solar cell semiconductor material, and the temperature of the cell.

4.3.1.6 *Short-circuit current*

The current through the solar cell is known as the short-circuit current when it is short-circuited and there is zero voltage drop across the solar cell. The short-circuit current is considered to be the light-generated current and it is the maximum current that can be drawn from a solar cell. In a practical solar cell, the short-circuit current is always less than the light-generated current because of the internal resistance of the solar cell. The short-circuit current is determined by Eq. (4.4), as shown below.

$$I = I_{Ph} - I_0\left[\left(e^{\frac{V_{sh}}{n.V_T}}\right) - 1\right] - \frac{V + IR_s}{R_{sh}}$$

4.3.1.7 The efficiency of a solar cell

The efficiency of the solar is defined as the ratio of the output energy that is being supplied by the solar cell to the energy available in the solar irradiance that is falling on the solar cell. The efficiency of the solar cell is always vulnerable to solar irradiance and the cell temperature, which we will discuss in detail. Mathematically, efficiency is determined by Eq. (4.7).

$$\eta = \frac{P_{max}}{P_{in}} = \frac{I_{max} \times V_{max}}{P_{in}} \tag{4.7}$$

By using the fill factor equation, P_{max} can be found as shown in the Eq. (4.8).

$$\eta = \frac{I_{sc}.V_{oc}.FF}{P_{in}} \tag{4.8}$$

4.3.1.8 Fill factor

The fill factor of a solar cell can be defined by Fig. 4.7. The area under the bigger rectangle indicates the maximum power that would be available from an ideal solar cell, whereas the area under the smaller rectangle represents the maximum power that a solar cell can deliver. Hence the fill factor is defined as the ratio of the maximum power that a solar cell can deliver to the maximum power that would be available from an ideal solar cell, as shown in Fig. 4.7. The fill factor is determined by Eq. (4.9).

$$FF = \frac{I_{max} \times V_{max}}{I_{sc} \times V_{oc}} \tag{4.9}$$

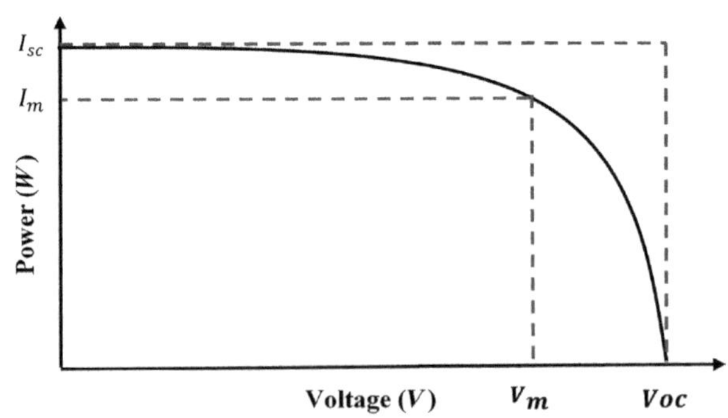

FIGURE 4.7 I–V curve for fill factor.

Example 4.1

A 100 cm^2 silicon solar cell operating at standard test conditions (STC) is giving 2.5 A short-circuit current. From the data sheet it was observed that the maximum voltage and maximum current were 0.6 V and 2.1 A. Find

1. The open-circuit voltage of the solar cell.
2. The fill factor of the solar cell.
3. The efficiency of the solar cell.
4. The maximum power point (MPP).

Given/required data

$$n = \text{ideality factor} = 1.3$$

$$k = \text{Boltzman constant} = 1.381 \times 10^{-23} J/K$$

$$T = 25^{\circ}\text{C}$$

$$q = \text{charge on electron} = 1.602 \times 10^{-19}\text{C}$$

$$V_m = 0.6V$$

$$I_m = 2.1A$$

Solution

1. Open-circuit voltage of the solar cell is determined by Eq. (4.6). Ideality factor, Boltzmann constant, and the charge on the electron are to be known from the literature.

$$V_{oc} = ln\left(\frac{I_{Ph}}{I_0} + 1\right)\left(\frac{n \cdot k \cdot T}{q}\right)$$

$$V_{oc} = ln\left(\frac{2.5}{10^{-12}} + 1\right)\left(\frac{(1.3)(1.381 \times 10^{-23})(298)}{1.602 \times 10^{-19}}\right)$$

$$V_{oc} = 0.94 \ V$$

2. Fill factor of the given solar cell is determined by Eq. (4.9)

$$FF = \frac{I_{max} \times V_{max}}{I_{sc} \times V_{oc}}$$

$$FF = \frac{2.1 \times 0.6}{2.5 \times 0.94}$$

$$FF = 0.54$$

3. Efficiency of the solar cell is determined by Eq. (4.8)

$$\eta = \frac{I_{sc}.V_{oc}.FF}{P_{in}}$$

$$\eta = \frac{2.5 \times 0.94 \times 0.54}{1000 \times 100 \times 10^{-4}} \times 100$$

$$\eta = 12.69\%$$

4. MPP of the solar panel is determined by multiplying voltage and current at the MPP.

$$P_{max} = I_{max} \times V_{max}$$
$$P_{max} = 2.1 \times 0.6$$

$$P_{max} = 1.26W$$

4.4 Effect of temperature on solar cell

The resistance of the solar cell is a parameter that directly influences the performance of the solar cell. At higher temperatures, more electrons vibrate and generate resistance to the flow of electrons to the external circuit. With an increase in temperature, the atoms start vibrating, which reduces the binding energy. The reduction in binding energy means that the electrons are less tightly bound in the valence band and they need a little energy to jump to the conduction band. This leads to the concept that the increasing temperature reduces the bandgap of the solar material. From the equation, it seems that the open-circuit voltage is directly proportional to the temperature but in reality, this is not true as reverse saturation current increases with the increase in temperature as depicted in the equation. Fig. 4.8 shows the P–V curves of the solar cell at different temperatures. As the temperature is increased from 25°C to 55°C in three steps, the open-circuit voltage and the maximum power of the solar cell are reduced. Fig. 4.9 shows the I–V curves

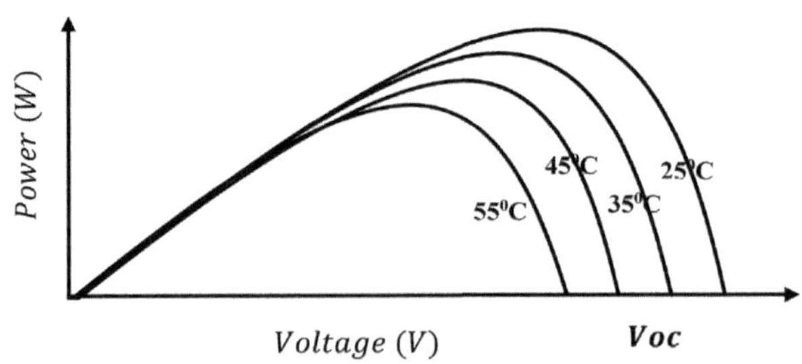

FIGURE 4.8 P–V curves at varying cell temperature.

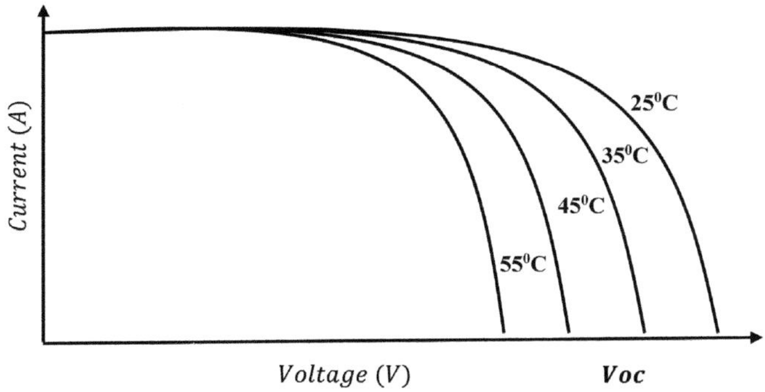

FIGURE 4.9 I–V curves at varying cell temperature.

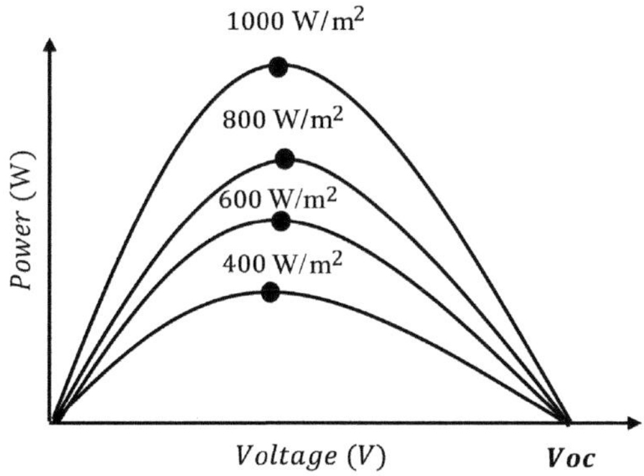

FIGURE 4.10 P–V curves at varying solar irradiance.

of the same solar cell at the same temperatures which also shows the same relation between temperature and the open-circuit voltage. The short-circuit current is not affected by temperature variations.

If an arrangement is made to filter out the radiation having less energy than the bandgap energy of the semiconductor material, the temperature of the solar cell can be controlled.

4.5 Effect of irradiance on solar cell

Because of the cloudy season and the absence of a solar tracker, the solar radiation directed to a solar cell is changing. Early in the morning, the solar cells are not directed toward the sun, which reduces the solar irradiance falling on the solar cells. Less solar irradiance means less photogenerated current. The equation shows that the photocurrent and the open-circuit voltage are directly proportional to each other. Fig. 4.10 shows the P–V curves at different solar irradiances. At 1000 W/m^2, the solar cell generates maximum

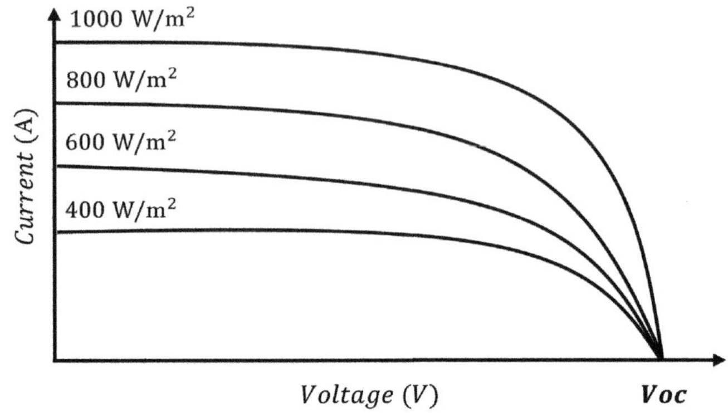

FIGURE 4.11 I–V curves at varying solar irradiance.

power. As solar irradiance is reduced, the power at the output also goes low, as shown in the figure. The open-circuit voltage at each solar irradiance level is the same. The I–V curves of the solar cell at different irradiance levels are shown in Fig. 4.11. The short-circuit current is directly proportional to the photocurrent which is directly proportional to the solar irradiance as given in the equation. The figure shows the increase in short-circuit current with an increase in irradiance and the open-circuit voltage remains the same. On comparing the figure and figure we can say that the increase in power due to an increase in irradiance is actually because of the increase in short-circuit current.

4.6 Series and parallel connection of solar cells

Series and parallel connections of low power solar cells are intended to increase the power. The parallel connection of solar cells increases the current output, whereas the series connection increases the voltage output. The characteristics of the solar cells being connected in series and parallel must be identical.

Consider a solar cell with I_{sc} of 0.6 V and V_{oc} of 0.4 V. The I–V characteristic curve of this solar cell is shown in Fig. 4.12.

For our requirement, if we need more voltages and the same current we will connect solar cells in series that will increase the voltage level keeping the current constant. If we connect two solar cells with the abovementioned characteristics in series, the open-circuit voltages of both cells will add up together to 0.8 V, while the current will sustain at 0.6 A. In Fig. 4.13 the gray line indicates the I–V curve of a single solar cell, whereas the black line indicates the I–V curve of two solar cells in series.

Similarly, for our required application, if we need the same voltages and increased current we will connect solar cells in parallel, which will increase the current keeping the voltage constant. If we connect two solar cells with

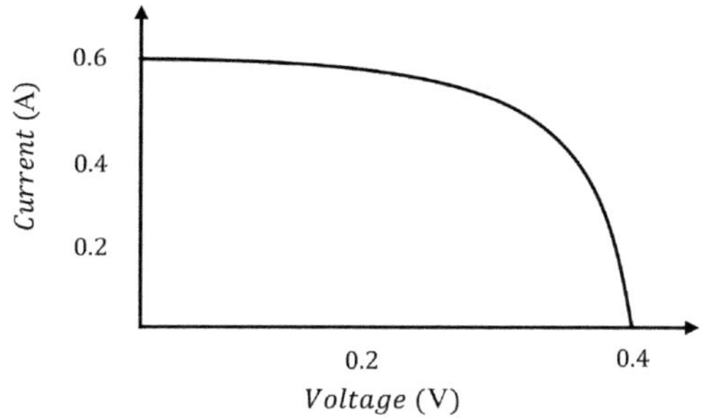

FIGURE 4.12 I–V curve of a single solar cell.

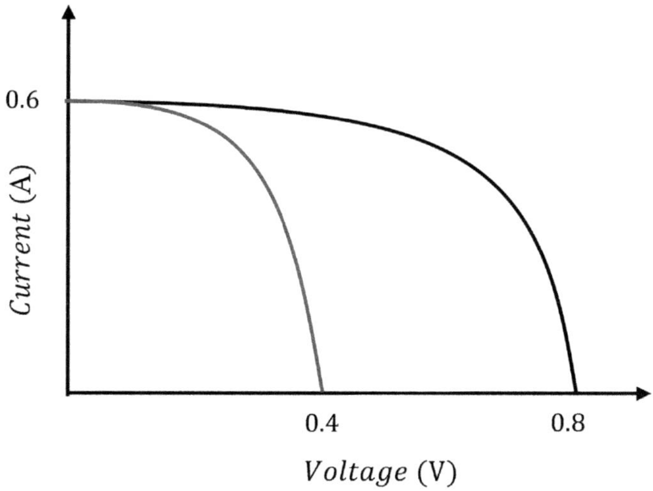

FIGURE 4.13 I–V curve of two series solar cells.

the abovementioned characteristics in parallel, the short-circuit current of both cells will add up together to 1.2 A, while the open-circuit voltage will sustain at 0.4 V. In Fig. 4.14, the gray line indicates the I–V curve of a single solar cell, whereas the black line indicates the I–V curve of two solar cells in parallel.

In contrast to the above two scenarios, if we need more voltage and more current, we will have to connect the solar cells in series and parallel combinations. If we connect two solar cells in series and then connect this series combination of solar cells with another series combination of solar cells in parallel, we will have added voltage and added current at the output. Fig. 4.15 shows the series and parallel combination of solar cells that increased the open-circuit voltage to 0.8 V and short-circuit current to 1.2 A. In Fig. 4.15, the gray line indicates the I–V curve of a single solar cell, whereas the black line indicates the I–V curve of two series solar cells in parallel to the other two series solar cells.

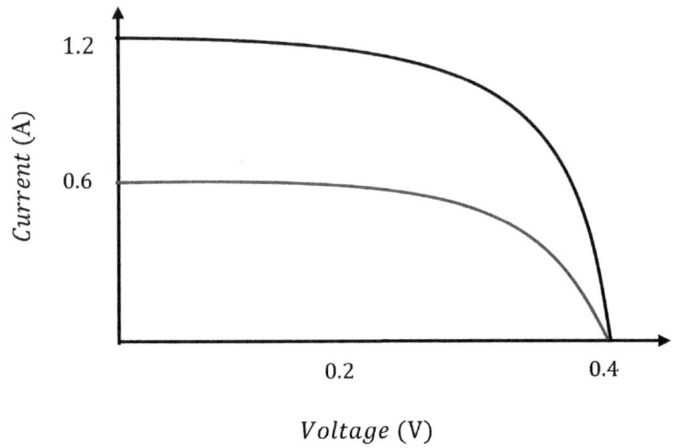

FIGURE 4.14 I–V curves of two parallel solar cells.

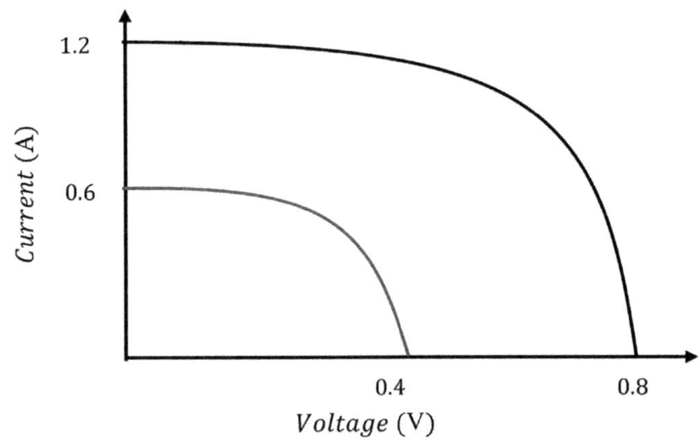

FIGURE 4.15 I–V curve of series and parallel combination of solar cells.

Example 4.2

Designing a solar PV system for a solar pump, you as a renewable energy engineer, determined that you needed to connect the PV panels in such a way that you get 24 V at the output with 15 A of current. You are required to connect the solar panels in such a way that you get the required voltage and current. The panels that you have are 12 V and 5 A.

Solution

As we have solar panels of 12 V each. We will have to connect two such solar panels in series to get 24 V.

$$\begin{aligned}\text{voltage in a single series branch} &= 12 + 12 \\ &= 24\,V\end{aligned}$$

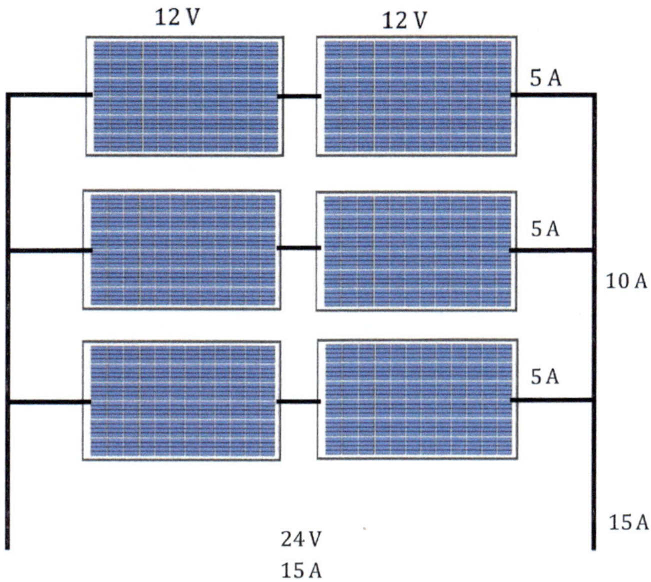

FIGURE 4.16 Series and parallel combination of solar cells.

Now, we have solar panels of 5 A current each. To get 15 A we will have to connect three series combinations of solar panels in parallel with each other to get 15 A.

$$\begin{aligned} \text{current through the circuit} &= 5 + 5 + 5 \\ &= 15\,A \end{aligned}$$

Hence, to get the 24 V with 15 A of current at the output, we will have to connect three parallel branches of two series solar panels, as shown in Fig. 4.16.

4.7 Solar tracker

Since the position of the sun in the sky is intermittent, the processes and the apparatuses that use the sunlight to process their function are necessary to be directed toward the sun the whole day. The solar PV system, solar parabolic trough, solar dish, solar tower, and ILS are some of the devices where solar trackers are used to properly function. Solar trackers could be single axis or dual axis based on the applications. In single-axis solar trackers only the east–west path in the sky is covered, whereas in dual-axis solar trackers the path of the sun, as well as the tilted angle north–south, is covered.

The position of the sun in the sky is described by two angles named the azimuthal angle and the altitude angle. The vertical line on the Earth's surface is called the zenith line and the angle between the zenith line and the

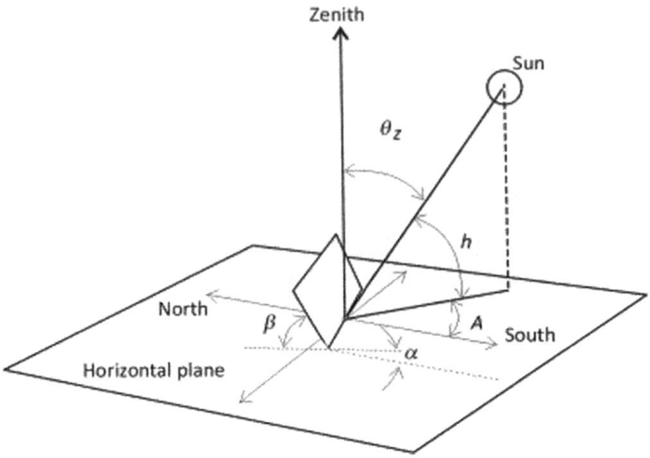

FIGURE 4.17 Zenith, altitude, and azimuthal angles.

line from the Earth's surface to the sun is called the zenith angle (θ_z). The complementary angle of the zenith angle is called the altitude angle (h). The azimuthal angle (A) is defined as the angular displacement from the south of the projection of the beam radiation on the horizontal plane. In normal condition A is zero. If A is greater than zero, the direction is toward the west. If it is less than zero or negative, the direction is east. The defined angles are described in Fig. 4.17.

Once the above-defined two angles are determined, the solar position and the amount of insolation reaching the point of the surface where the solar panels are intended to be installed can be exactly found.

4.7.1 Single-axis solar tracker

In single-axis solar trackers for solar PV, the sun is tracked from east to west to orient the solar panels toward the sun throughout the day. This tracking maximizes solar output. In single-axis solar tracker, a motor is used to rotate the panel mounting structure. The decision to rotate the motor either toward east or west is taken by comparing the output of the light-dependent resistors LDR_E and LDR_W. The figure shows the positioning of the LDR_E and LDR_W before and after tracking. Fig. 4.18A shows the unequal collection of solar irradiance by the LDR_E and LDR_W. LDR_E is in shadow getting lower irradiance, whereas LDR_W is getting maximum solar irradiance. So, the solar panel is not directed toward the sun. To get it oriented toward the sun, it is necessary for both the *LDR*s to get equal solar irradiance. Fig. 4.18B shows the *LDR*s after tracking when both the *LDR*s are receiving equal solar irradiance.

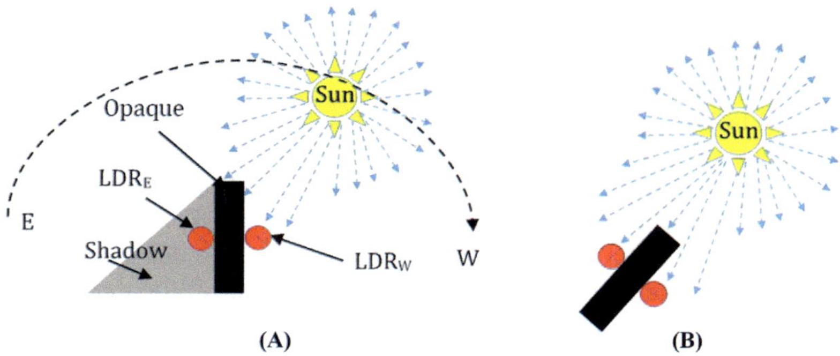

FIGURE 4.18 Solar tracker: (A) *LDR* before tracking; (B) *LDR* after tracking.

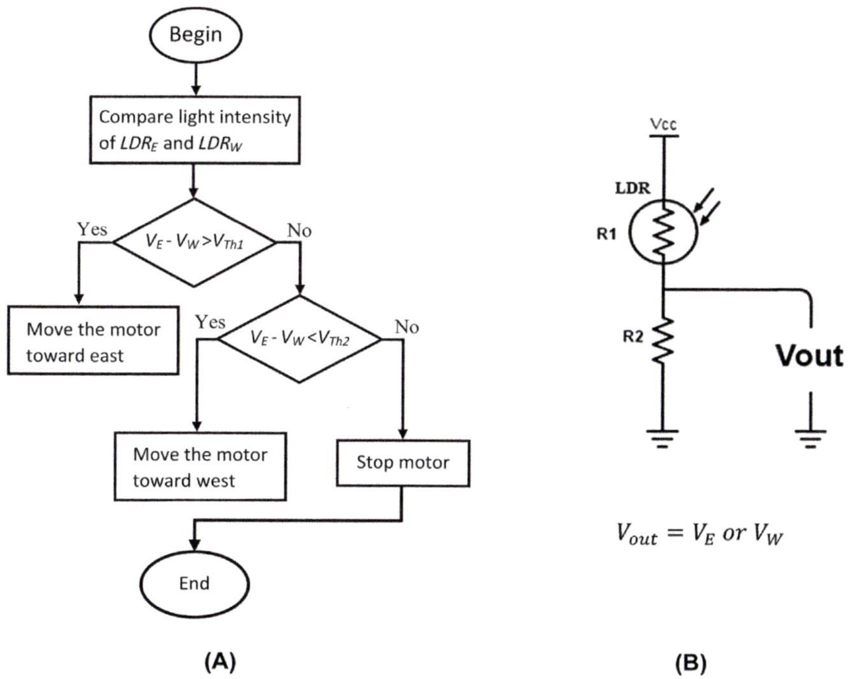

FIGURE 4.19 (A) Flow chart of single-axis solar tracker. (B) Solar irradiance measuring technique.

As the solar irradiance falling on the *LDR* decreases, its resistance is increased. An *LDR* is connected to a series resistor and by applying the voltage dividing rule the voltage across the series resistor is measured as shown in Fig. 4.19B. This is done for both LDR_E and LDR_W and the output voltage is measured. According to the flow chart shown in Fig. 4.19B, the decision is made as to whether to rotate the motor toward the east or west.

After taking a decision and rotating the motor toward the west we can see in Fig. 4.19B that now both the LDR_E and $LDR_W$are getting equal irradiance and the panel is directed toward the sun.

Fig. 4.19A shows the flow chart of the single-axis solar tracker where the voltage ($V_E$andV_W) from two LDR apparatuses is compared. As in Fig. 4.18A V_E is greater than V_W indicating that the LDR_E is in the shade so it must be brought in sunlight by rotating the motor toward the west, as shown in Fig. 4.18B. If V_E is equal to V_W, it indicates that both *LDR*s are receiving the same irradiance and we don't need to move the motor in either direction. This single-axis solar tracker covers the altitude angle of the solar panel.

4.7.2 Dual-axis solar tracker

Dual-axis solar trackers are used where more accuracy is required in directing the solar panels toward the sun. Here the solar panel is moved in both altitude direction and the zenith angle. The working principle of the dual-axis solar tracker is the same as that of the single-axis solar tracker. The flow chart of the dual-axis solar tracker is shown in Fig. 4.20. Here first the path of the sun from east to west is tracked using a single motor and then the tilt in the north–south direction is tracked.

4.8 Maximum power point tracker

4.8.1 Perturb and observe

To derive the Perturb and Observe (P&O) MPPT algorithm, the operating point of the PV cell at the four different points of the P–V curve is observed and, accordingly, perturbation is made in the operating voltage of the solar cell. Fig. 4.21 shows the first case where the operating voltage at two different consecutive times $T(k)$ and $T(k-1)$ is observed, named as $V(k)$ and $V(k-1)$, respectively [4]. The observation indicates that the operating voltage is to the right of the MPP and is decreased. To reach the MPP, the operating voltage needs to be further decreased until it reaches the MPP by the perturbation in the operating voltage. The perturbation in the voltage is done by a perturbation in the duty cycle of the converter coupled to the solar cell. In the present case, the duty cycle is decreased to decrease the voltage and hence to reach the MPP [5].

Fig. 4.22 shows the second case where the operating voltage at two different consecutive times $T(k)$ and $T(k-1)$ is observed, named as $V(k)$ and $V(k-1)$, respectively. The observation indicates that the operating voltage is to the left of the MPP and being increased. To reach the MPP, the operating voltage needs to be further increased until it reaches the MPP by the

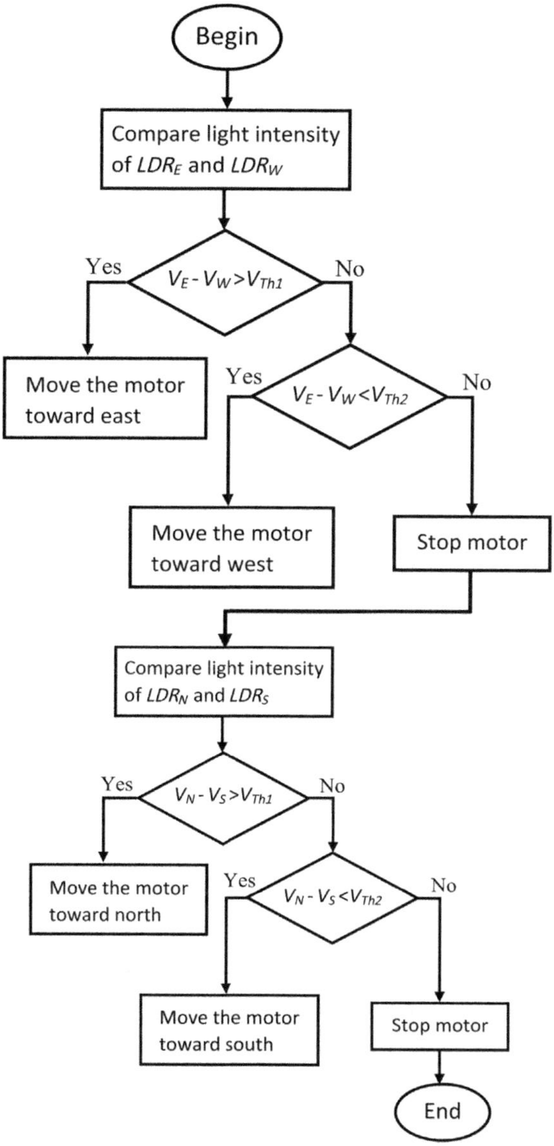

FIGURE 4.20 Flow chart of dual-axis solar tracker.

perturbation in the operating voltage. In the present case, the duty cycle is increased to increase the voltage and hence to reach the MPP.

Fig. 4.23 shows the third case where the operating voltage at two different consecutive times $T(k)$ and $T(k-1)$ is observed, named as $V(k)$ and $V(k-1)$, respectively. The observation indicates that the operating voltage

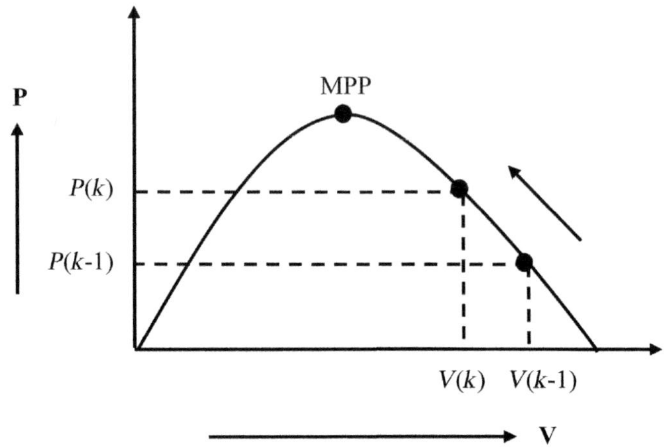

FIGURE 4.21 Case 1, when operating point approaches MPP from right.

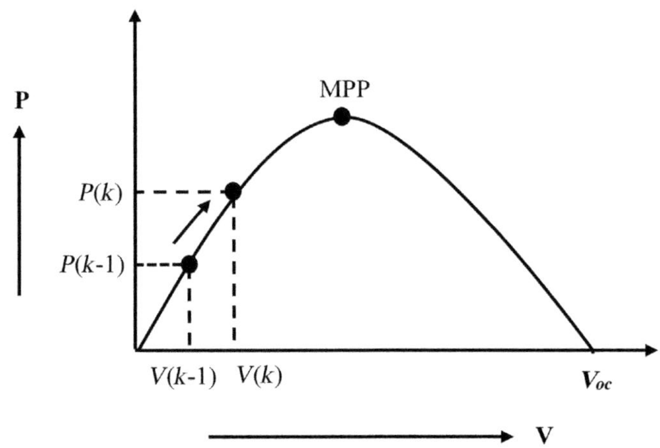

FIGURE 4.22 Case 2, when operating point approaches MPP from left.

is to the left of the MPP and is decreased. To reach the MPP, the operating voltage needs to be increased in the reverse direction until it reaches the MPP by the perturbation in the operating voltage. In the present case, the duty cycle is increased to increase the voltage and hence to reach the MPP.

Fig. 4.24 shows the fouth case where the operating voltage at two different consecutive times $T(k)$ and $T(k-1)$ is observed, named as $V(k)$ and $V(k-1)$, respectively. The observation indicates that the operating voltage is to the right of the MPP and being increased. To reach the MPP, the operating voltage needs to be decreased in the reverse direction until it reaches the MPP by the perturbation in the operating voltage. In the present case, the

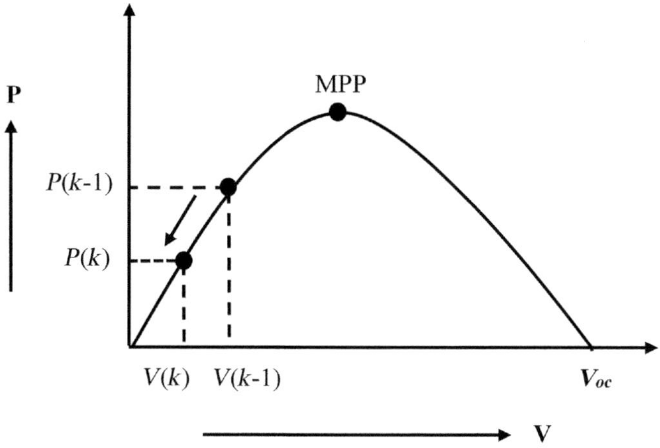

FIGURE 4.23 Case 3, when operating point leaves the MPP from left.

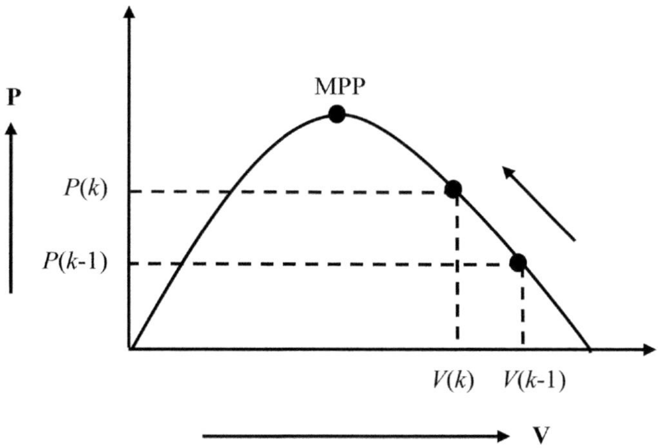

FIGURE 4.24 Case 4, when operating point leaves the MPP from right.

duty cycle is decreased to decrease the voltage and hence to reach the MPP [5]. Based on these four cases, a flow chart of the P&O algorithm is derived, as shown in Fig. 4.25. A summary of the P&O algorithm is given in Table 4.2.

4.8.2 Incremental conductance

An incremental conductance (INC) algorithm is derived by using the power formula of a DC source [6].

$$P = VI \tag{4.10}$$

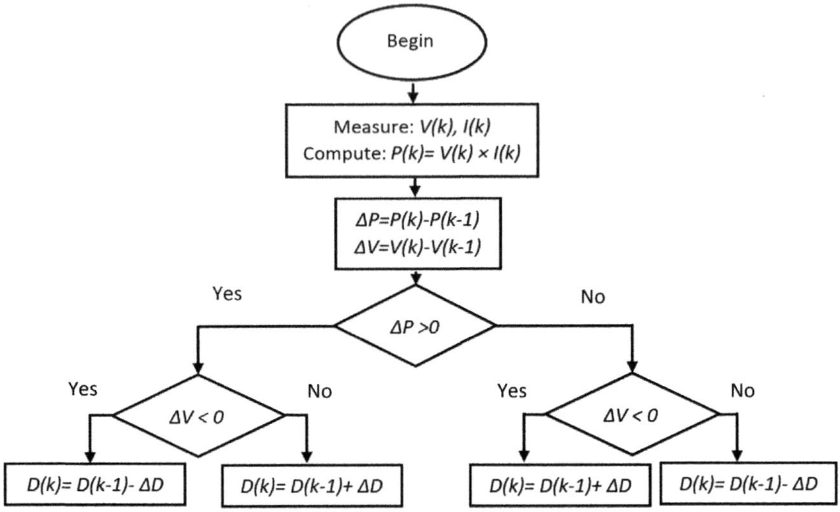

FIGURE 4.25 Flow chart of P&O MPPT.

TABLE 4.2 Summary of P&O algorithm.

Case	Observation	Perturbation
Case 1	Power increases $\Rightarrow \Delta P > 0$	ΔD subtracted
	Voltage decreases $\Rightarrow \Delta V < 0$	
Case 2	Power increases $\Rightarrow \Delta P > 0$	ΔD added
	Voltage increases $\Rightarrow \Delta V > 0$	
Case 3	Power decreases $\Rightarrow \Delta P < 0$	ΔD added
	Voltage decreases $\Rightarrow \Delta V < 0$	
Case 4	Power decreases $\Rightarrow \Delta P < 0$	ΔD subtracted
	Voltage increases $\Rightarrow \Delta V > 0$	

Differentiating both sides of Eq. (4.10) with respect to V.

$$\frac{dP}{dV} = \frac{d(VI)}{dV}$$

$$\frac{dP}{dV} = V \times \frac{dI}{dV} + I \tag{4.11}$$

To find the relation at the maximum power, we put first derivative (slope of the P–V curve) equal to zero in Eq. (4.11) as the slope of the P–V curve at the MPP is zero, as shown in Fig. 4.26. Eq. (4.12) is the relation between

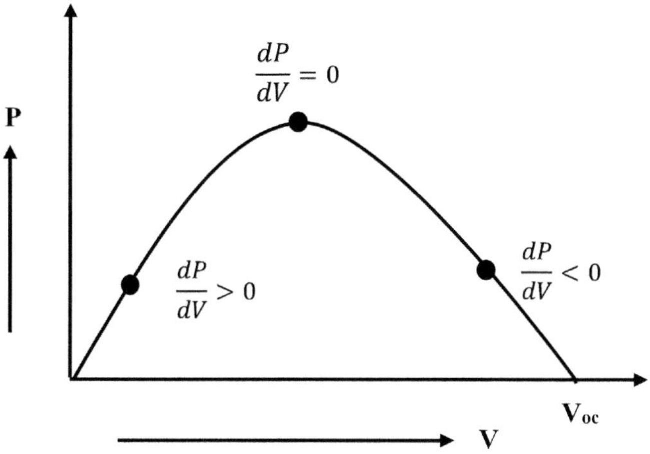

FIGURE 4.26 Slopes of the P–V curve.

conductance and the INC for the MPP. Similarly by using the slopes of the P–V curve at the left of MPP and the right of MPP we can determine Eqs. (4.13) and (4.14). Now with the help of Eqs. (4.12)–(4.14), we can draw the flow chart of the INC algorithm shown in Fig. 4.27.

$$-\frac{I}{V} = \frac{dI}{dV} \tag{4.12}$$

$$-\frac{I}{V} > \frac{dI}{dV} \tag{4.13}$$

$$-\frac{I}{V} < \frac{dI}{dV} \tag{4.14}$$

4.9 Off-grid PV system

Among renewable energy sources, solar PV systems are widely used in both off-grid and on-grid mode. Direct-coupled DC load to the PV system is the simplest configuration of the PV system, mostly used in areas far away from the main electricity grid. For small applications, a solar panel is directly connected to the DC load. The direct-coupled PV system configuration is shown in Fig. 4.28. As this configuration is batteryless, the load is entertained only in sunlight hours. The variation in sun irradiance throughout the day causes a speed variation in the DC fan, dim lighting in LED lights, variation in water flow in DC pumping, and other loads [7].

The other regulated configuration of the off-grid PV system consists of battery storage, charge controller, and inverter, as shown in Fig. 4.29. Depending on the state of the charge (SOC) of the battery, the charge

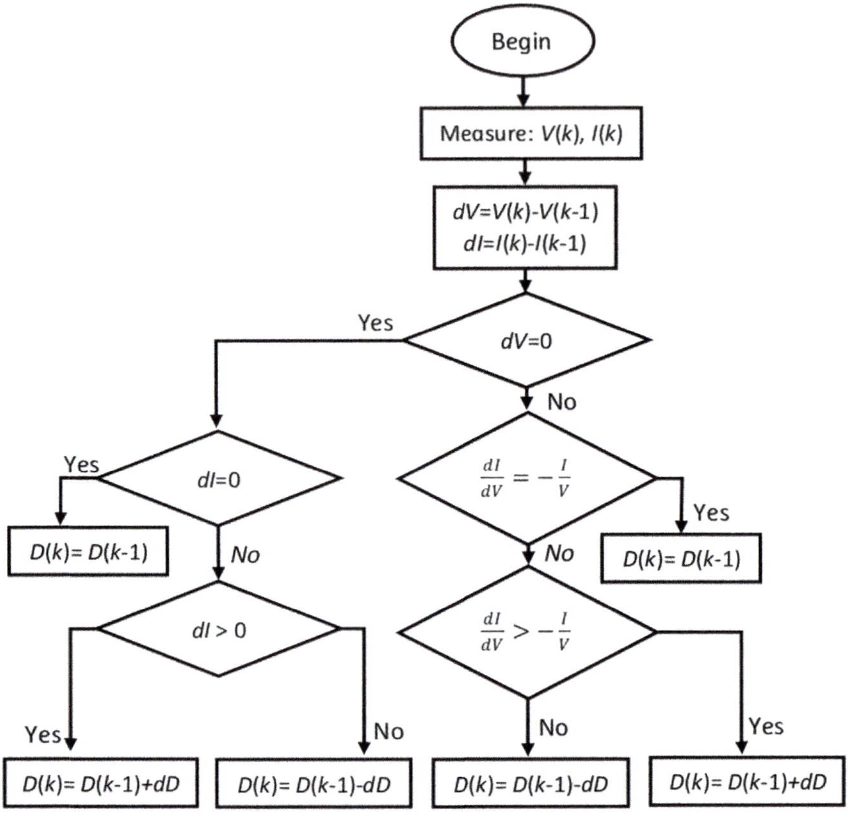

FIGURE 4.27 Flow chart of incremental conductance algorithm.

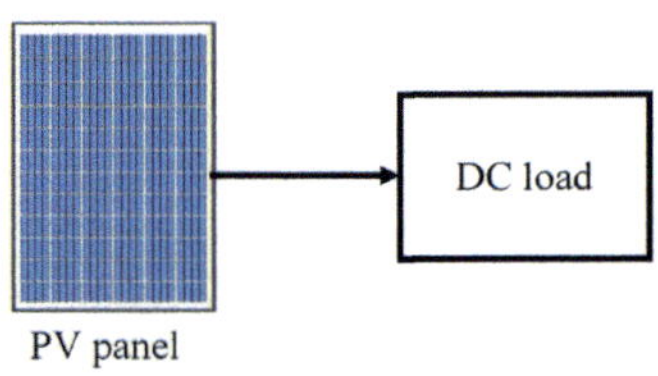

FIGURE 4.28 Unregulated solar photovoltaic system with direct current load.

controller decides whether to charge the battery or entertain the DC load. The charge controller maintains control over the deep discharging and overcharging. In both cases, effectiveness and the lifetime of the battery are affected. The use of batteries increases the cost of the system. On priority bases, the SOC of the battery is maintained at its maximum value to be used in low insolation periods and at night when there is no enough irradiance to generate the electrons in the PV material. To entertain the AC loads, an inverter is used that takes the power from the batteries and converts the DC

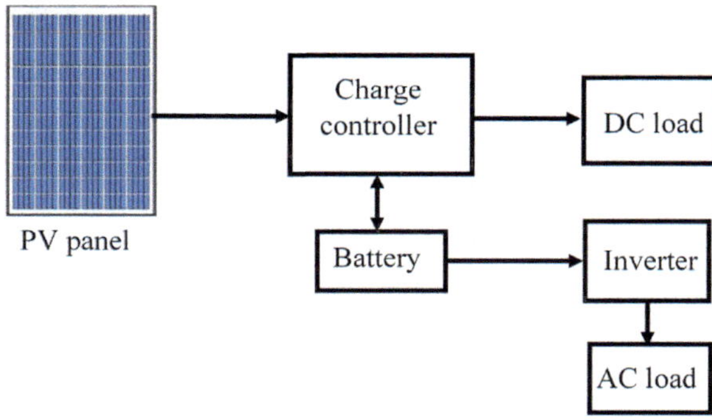

FIGURE 4.29 Regulated solar photovoltaic system with battery storage.

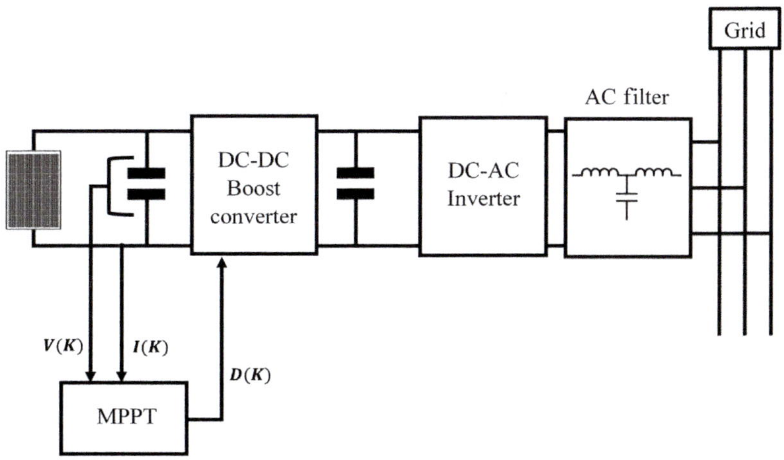

FIGURE 4.30 Grid-connected solar photovoltaic system.

to AC for AC loads. In some cases, to let the solar PV to operate at its MPP, a DC to DC converter is used, controlled by the MPP Tracker (MPPT).

4.10 Grid-connected PV system

As compared to the off-grid PV systems, no battery is needed in the grid-connected PV systems that reduce the cost of the system. The grid is considered as an infinite sink and source of electricity. In Fig. 4.30, a two-stage grid-connected PV system is shown. It consists of PV panels, a MPPT-based DC-DC boost or a buck-boost converter, a bidirectional inverter, and intermediate filter stages. In a grid-connected PV system, the surplus energy from the solar PV system during the peak and off-peak hours is sold to the grid. During low irradiance when there is less production of electricity from the solar PV system, the load requirement is fulfilled by importing electricity from the grid.

The efficiency of the PV system is determined by Eq. (4.15) as follows:

$$\eta_{PVsystem} = \eta_{PVmodules} \times \eta_{MPPTconv.} \times \eta_{inverter} \tag{4.15}$$

With the increased penetration of the PVs in the grid, the grid may have the following negative impacts.

1. Voltage control
2. Over voltage
3. Reverse power flow
4. Power quality
5. Phase unbalance
6. Islanding detection
7. Increased reactive power.

4.11 Hybrid PV systems

Because of the intermittent nature of renewable energy sources, a hybrid renewable energy system is introduced. The various renewable energy sources integrated are well recognized to entertain the load in the locality not connected to the grid. Any of the sources like solar, wind, geothermal, fuel cell, micro hydro, and biogas can be integrated with auxiliary components. The resulting system is termed the hybrid renewable energy system. Apart from renewable energy sources, conventional power plants can also be integrated to meet the load during the lesser- or nonavailability of renewable energy sources. Solar-based hybrid energy systems can be configured in three following ways [8].

4.11.1 Series hybrid energy system

A series PV hybrid energy system integrated with a diesel generator is shown in Fig. 4.31. The output of the diesel generator is rectified using an AC to

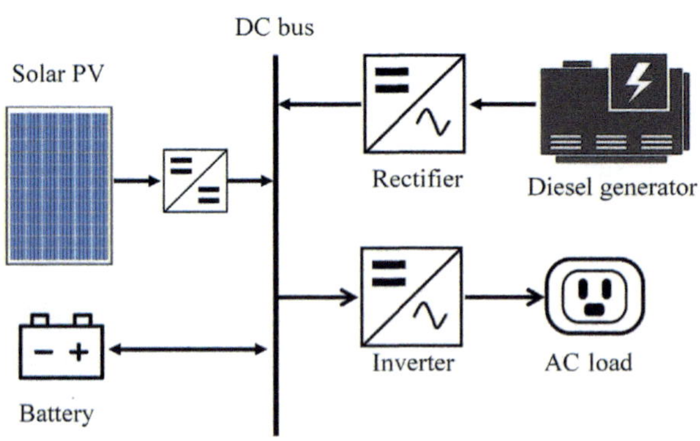

FIGURE 4.31 Series photovoltaic hybrid energy system.

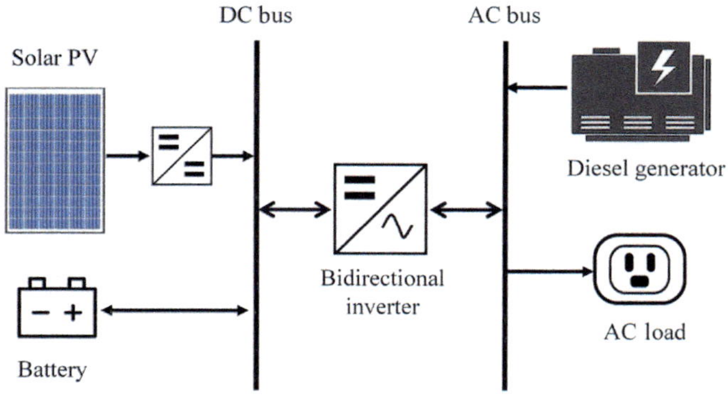

FIGURE 4.32 Parallel photovoltaic hybrid energy system.

DC rectifier and dumped on the DC bus. The output of the solar PV is also dumped on the DC bus through a DC-DC boost converter. The DC-DC boost converter also ensures the MPPT to allow the PV panels to operate at maximum power. A battery is also connected to the DC bus that entertains the load during low irradiance. The inverter connected to the DC bus converts the DC into AC to entertain the AC loads. This configuration is termed as series configuration as the diesel generator is not directly connected to the inverter. One disadvantage of this configuration is the losses in the rectifier.

4.11.2 Parallel hybrid energy system

Contrary to the series hybrid energy system, AC and DC buses are present in the parallel hybrid energy system, as shown in Fig. 4.32. The diesel generator and the inverter are in parallel to the AC load. A bidirectional inverter is used to link the diesel generator and the battery bank. A bidirectional inverter acts as a rectifier to charge the battery if the energy from the diesel generator and the solar PV is greater than the load required. So here the inverter of less capacity is required as compared to the series hybrid energy system since it has to invert the voltage of solar PV only. However, in this configuration, synchronization of the diesel generator and the solar PV is required.

4.11.3 Switched hybrid energy system

In a switched hybrid energy system, the load is either entertained by the diesel generator or the inverter. The battery is charged either by the diesel generator or solar PV. The major advantage of this configuration is that the diesel generator can directly entertain the load without bearing any conversion losses. If the load demand is less than the generation of the diesel generator, the surplus energy will be used to charge the battery. During off-peak hours, the diesel generator is switched off and the load is entertained by the solar PV and the

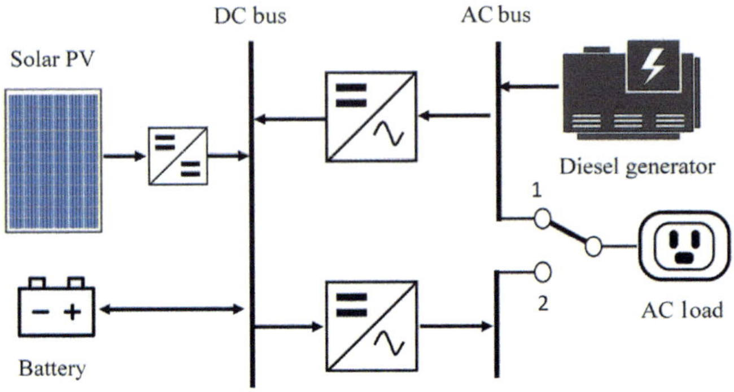

FIGURE 4.33 Switched hybrid energy system.

battery storage. In both modes, the efficiency of the system is increased. The configuration of the switched hybrid energy system is shown in Fig. 4.33.

4.12 Distributed generation

The conventional energy sources like coal, gas, and oil are rapidly depleting while the world energy demand is growing more rapidly. A diversity of renewable energy sources will replace conventional energy sources and the conventional transmission and distribution system. The renewable energy sources being abundant creates the idea of distributed power generation [9]. Distributed power generation is the generation of electricity at the consumer side, distribution feeders, or the substations by the locally installed wind, solar, fuel cell, biomass, and other sources. DG can integrate both conventional energy sources and renewable energy sources [10]. Wind energy, solar energy, and micro hydro energy are the most promising renewable energy sources in distributed generation (DG). Fig. 4.34 shows DG consisting of conventional and renewable energy sources, distributed load, and the distribution grid. Fig. 4.34 also shows the integration of the distribution grid to the transmission grid [11].

A major component of the DG is net metering. The customers generating their electricity using solar, wind, or biogas plants can sell the energy beyond their needs to the grid and can purchase the electricity from the grid when their production is lower than their load requirement. In other words, the customers store their electricity on the grid and use it later in peak hours avoiding other extra charges.

4.13 Optimization of hybrid renewable energy system

With the increasing energy demands and the depleting conventional energy sources, demand and innovation of the renewable energy sources has

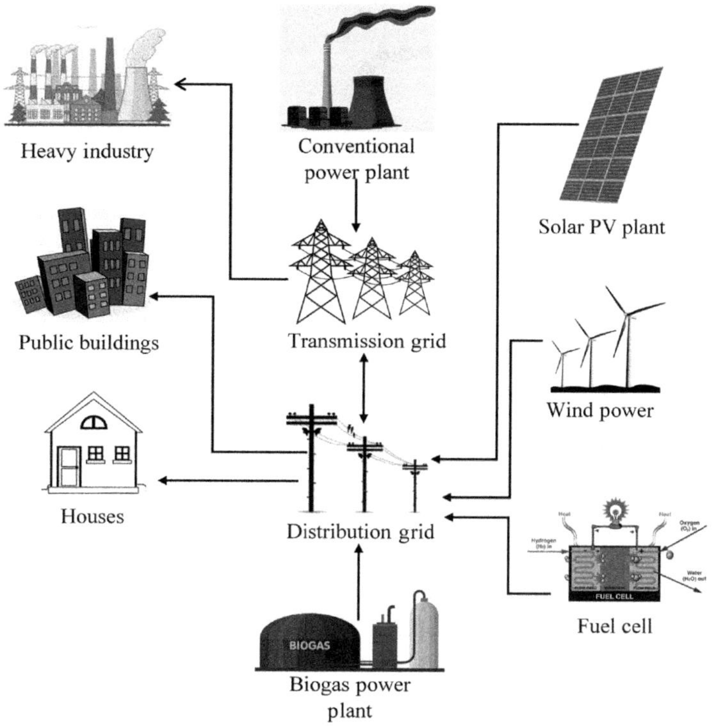

FIGURE 4.34 Distributed generation.

increased. As the renewable energy sources are intermittent and are available at different locations, hybrid renewable energy systems are proposed. To meet the required load, optimization of the hybrid energy system is performed to minimize the levelized cost of energy.

4.13.1 HOMER pro

HOMER software (Hybrid Optimization Model for Renewable Energy System) is used for the optimization of the renewable energy systems. The optimization process in HOMER is shown in Fig. 4.35. Initially the problem is assessed and the site of the project is decided with the load demanded, resources available, and the feasibility of the proposed design of the system. Hourly load data, monthly resources available, and the economics of the components are provided to HOMER as inputs. HOMER takes the coordinates of the location where the project installment is to be intended and fetches the wind speed, solar global horizontal irradiance (GHI), DNI, temperature, and the humidity data from NASA.

While designing a power system, the designer has to take various decisions like which components, how many components, and what size would

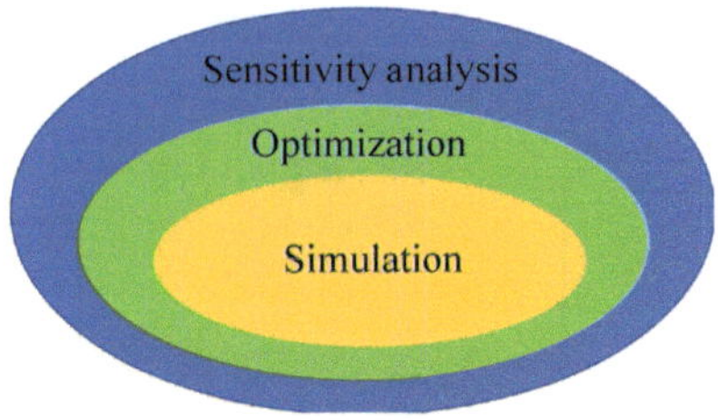

FIGURE 4.35 Onion diagram of the process of HOMER.

be the best for the hybrid energy system? A huge variation in the cost, intermittent sources of energy and the available technologies make the choice difficult. The optimization and sensitivity analysis of the HOMER make it easy to make the decision. HOMER models DG systems for both on-grid and off-grid. For ease of understanding the working of HOMER, it is necessary for its three core functionings to be understood, as described in the onion diagram shown in Fig. 4.35.

- Simulation: HOMER simulates all the possible combinations of the components, energy sources, and economics that you want to use in your hybrid model. HOMER simulates thousands of systems considering the above parameters.
- Optimization: all the simulated models are optimized based on the predefined criteria by the user that could be either COE, capital cost, net present value, and many others. After optimizing all the possible simulations, HOMER gives the best optimal solution.
- Sensitivity analysis: HOMER provides an optional step that allows the user to use the various values of the variable that are vulnerable to change from time to time like fuel cost, solar irradiance, and the wind speed. HOMER simulates the system for every change in the input variable and sees how the results vary for every variation in the input variable. HOMER contains the following components that are used in modeling and simulating a hybrid renewable energy system.

4.13.2 iHOGA

Improved Hybrid Optimization using Genetic Algorithms (iHOGA) was developed by the University of Zaragoza, Spain for optimum sizing of the hybrid renewable energy systems. iHOGA uses a double genetic algorithm. The first algorithm is used for the system components while the second one for the control strategy. The software can optimize a single objective and a multi objective system for off-grid and grid-connected renewable energy systems.

4.13.3 Hybrid2

Hybrid2 is an energy management software developed by the National Renewable Energy Laboratory to help the energy developers in conducting the preliminary study of the energy projects. Hybrid2 consists of four basic parts: Graphical User Interface (GUI), Simulation module, Economics module, and the Graphical Result Interface (GRI). The GUI allows the users to add the components to the system and the concerned inputs to these components. The simulation module allows the users to run the simulations with checking of the errors in the input variable. The economics module performs the economic analysis on the results generated by the Simulation module. The GRI allows the users to view the detailed outputs of the technical and economic analysis of the project graphically. Hybrid2 uses two types of simulation models. One is for long-term performance predictions, known as the logistic model. The other simulation model is for the rapid changes in the input variable, like wind speed solar irradiance etc., known as the dynamic model.

4.13.4 RETScreen

RETScreen is an Excel-based clean energy management software developed by the Natural Resources, Government of Canada. RETScreen helps in decision-making regarding the technical and economic feasibility of the project. Like HOMER, RETScreen takes the technical and economics input data and presents the emission reduction analysis and the financial analysis of the proposed project.

4.13.5 TRNSYS

TRANSYS is a graphical software developed to simulate the behavior of the transient systems. It can simulate the thermal systems, electrical energy systems and other dynamic systems like traffic flow system and the biological systems. TRANSYS consists of two parts: a kernel and a library of components. The kernel reads the input data, solves the system, finds the convergence, and plots the variables. The library of components contains almost 150 editable models like pump, electrolyzers, wind turbine, PV, and weather data processor which can be edited and modified by the user according to their own requirements.

4.14 Optimization of a hybrid energy system in HOMER: a case study

A hybrid renewable energy system as a case study is simulated here for the purpose of optimization of the best possible combination of the energy

TABLE 4.3 Components available in HOMER.

Power sources in HOMER	Storage in HOMER	Loads in HOMER
• Wind turbine • Solar PV • Fuel cell • Diesel generator • Run-of-river hydro • Hydro power • Biomass power • Microturbine • Electric utility grid	• Flywheels • Flow batteries • Hydrogen • Customizable batteries	• Daily profiles with seasonal variation • Thermal (space heating, crop drying) • Deferrable (water pumping, refrigeration) • Efficiency measures

sources, storage system, and the load. The objective function is to minimize the net present cost (NPC) of the hybrid energy system consisting of the hydro power, solar PV, wind turbine, diesel generator, converter, battery backup, and the load among the components, shown in Table 4.3.

The optimization of the presented hybrid renewable energy system is carried systematically in a process shown in Fig. 4.36. In pre-HOMER assessment, all the used conventional and renewable energy sources are estimated and given as the input to HOMER, load is calculated and also given as the input to HOMER. The third thing is the design considerations of the proposed system (Fig. 4.37).

4.14.1 Load assessment

The load of the location where the electricity from the proposed hybrid energy system is to feed is calculated by surveying the locality. In this case we are considering a microgrid consisting of five small houses, three large houses, two department stores, and two schools. Assessing the appliances of each home, schools, department stores, the load calculated is shown in Table 4.4.

4.14.2 Resource assessment

4.14.2.1 Solar photovoltaic

As described earlier, the proposed system consists of the solar PV, wind, and micro hydro renewable energy sources. HOMER takes the data for these resources on the given coordinates from the NASA surface meteorology and it can also import recorded values from a file. HOMER presents these resources' data in tabular and graphical forms. HOMER uses the solar

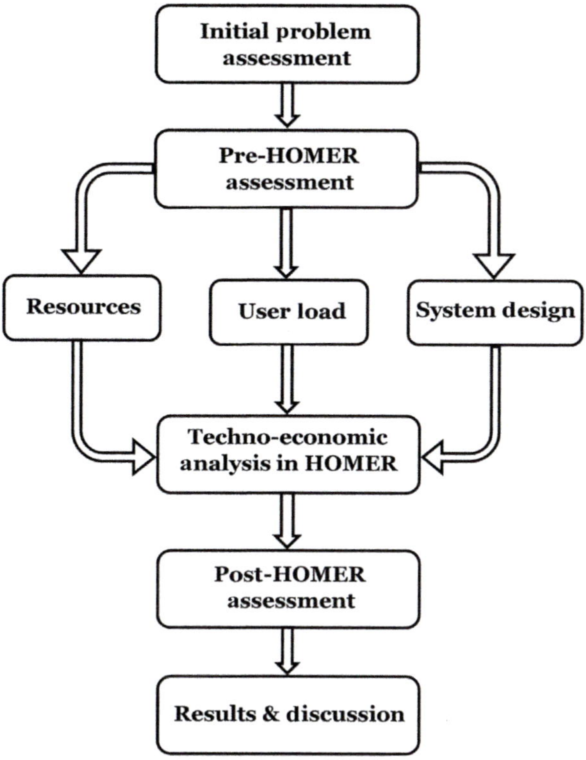

FIGURE 4.36 Optimization process of the Hybrid Renewable Energy System (HRES).

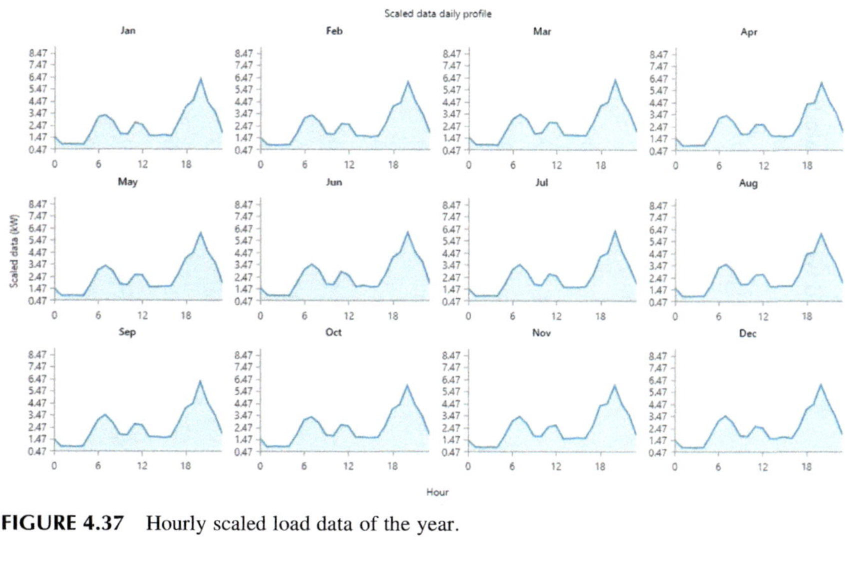

FIGURE 4.37 Hourly scaled load data of the year.

TABLE 4.4 Load assessment.

	Load type	Rated power (W)	Quantity	Hours	Energy (Wh/day)	Total energy (kWh/day)	No. of houses	Total (kWh/day)
Small/medium houses	Refrigerator	50	1	12	600			16
	TV	80	1	8	640			
	Light	20	3	8	480	3.32	5	
	Fan	80	2	14	1120			
	Miscellaneous	20	1	24	480			
Large houses	Electric motor	820	1	1	820			22
	TV	80	1	8	640			
	Light	20	5	8	800	7.40	3	
	Fan	80	3	14	3360			
	Refrigerator	100	1	13	1300			
	Miscellaneous	20	1	24	480			
Stores	Refrigerator	100	1	12	1200			7
	Light	20	4	5	400	3.36	2	
	Fan	80	1	16	1280			
	Miscellaneous	20	1	24	480			
Schools	Light	20	10	7	1400			15
	Fan	80	10	7	5600	7.48	2	
	Miscellaneous	20	1	24	480			
Total (kWh/day)								60

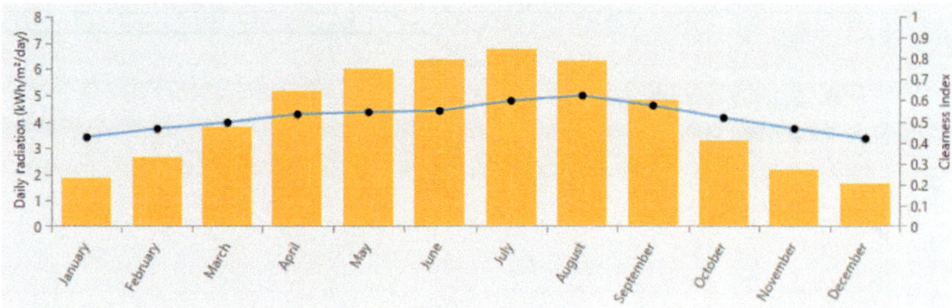

FIGURE 4.38 Solar global horizontal irradiance.

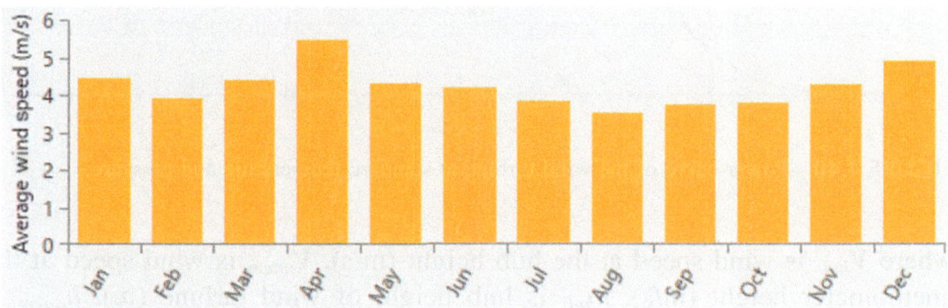

FIGURE 4.39 Wind speed throughout the year.

irradiance data as the input shown in the figure and calculates the output power from the solar PV using the formula given in Eq. (4.16).

$$P_{PV} = Y_{PV} f_{PV} . \left[\frac{\overline{G}_t}{\overline{G}_{t,STC}} \right] . \left[1 + \alpha_p \left(T_C - T_{C,STC} \right) \right] \tag{4.16}$$

where P_{PV} is power from the solar PV, Y_{PV} is power from the solar PV at STC (25°C, 1000 W/m^2), f_{PV} is derating factor of PV (%), $\overline{G}_t$ is solar irradiance on the PV cell at the current time T, $\overline{G}_{t,STC}$ is solar irradiance on the PV cell at STC (25°C, 1000 W/m^2), α_p is temperature coefficient of power, T_C is temperature of the PV cell at current time T (°C), and $T_{C,STC}$ is temperature of the PV cell at STC (25°C, 1000 W/m^2) (Fig. 4.38).

4.14.2.2 Wind power

HOMER calculates the output power from the wind turbine in different steps. In the first step, wind speed data given in Fig. 4.39 is at 10 m anemometer height. HOMER uses this data and scales it to the hub height by using either the power law or the algorithmic law, as specified by the user on the wind resource page using Eqs. (4.17) and (4.18), respectively.

$$V_{hub} = V_{anem} \left[\frac{\ln \left(h_{hub} / h_0 \right)}{\ln \left(h_{anem} / h_0 \right)} \right] \tag{4.17}$$

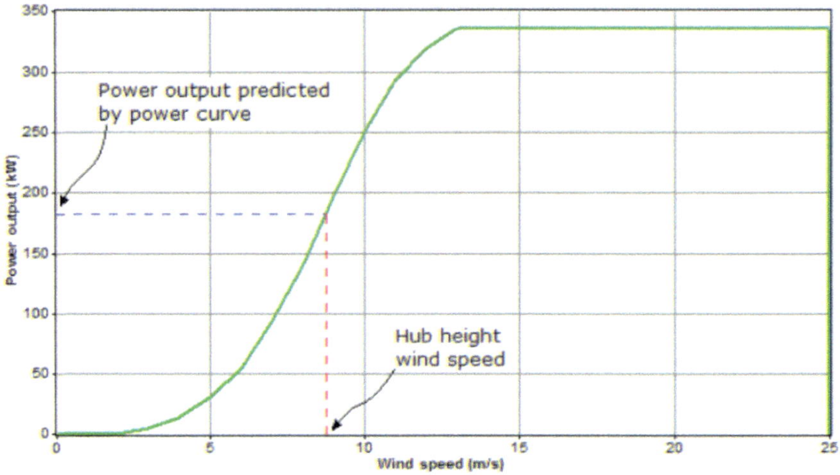

FIGURE 4.40 Power curve of the wind turbine at standard temperature and pressure.

where V_{hub} is wind speed at the hub height (m/s), V_{anem} is wind speed at the anemometer height (m/s), h_{hub} is hub height of wind turbine (m), h_{anem} is anemometer height (m), and h_0 is surface roughness length (m).

If we select the power law for the calculation of wind speed at the hub height, HOMER uses the power law given in Eq. (4.18).

$$V_{hub} = V_{anem}\left[\frac{h_{hub}}{h_{anem}}\right]^{\alpha} \tag{4.18}$$

where α is the exponent of the power law.

After calculating the wind speed at the height of the wind turbine or the hub height, HOMER uses the power curve of the turbine to determine the output power of that turbine at the wind speed at standard temperature and pressure conditions. The power curve of the turbine is the graph between output power of the turbine and the wind speed shown in the figure. The red line indicates the wind speed and the blue line indicates the corresponding output power of the wind turbine on the power curve. If the wind speed is beyond the power curve it indicates that the turbine does not produce any power below the minimum cutoff and above the maximum cutoff.

Fig. 4.40 shows the power curve at standard temperature and pressure, which is not possible at every wind site. HOMER uses Eq. (4.19) to perform the density correction by multiplying the power at STP to the density ratio to calculate the actual output power from the wind turbine.

$$P_{WT} = P_{WT,STP}\left(\frac{\rho}{\rho_0}\right) \tag{4.19}$$

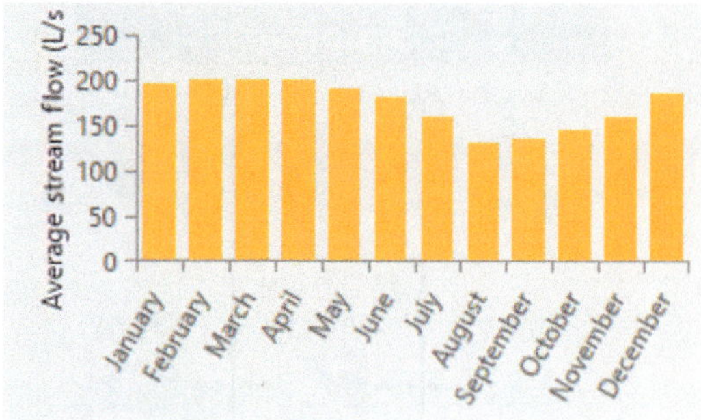

FIGURE 4.41 Flow rate throughout the year.

where P_{WT} is actual power from the wind turbine (kW), $P_{WT,STP}$ is power from the wind turbine at STP (kW), ρ is actual air density (kg/m^3), and ρ_0 is air density at STP. Power curve of the wind turbine at standard temperature and pressure is shown in Fig. 4.40.

4.14.2.3 Hydro energy

Hydro resources at the proposed location are shown in Fig. 4.41. HOMER asks about the other information on the hydro resource page like the available head, head losses, net head, density of the water, flow rate, efficiency of the turbine, and the coupled generator and uses Eq. (4.20) to calculate the power from the hydro turbine.

$$P_{hyd} = \rho \cdot g \cdot H_{net} \cdot Q \cdot \eta \tag{4.20}$$

where P_{hyd} is actual available power from hydro, ρ is water density, g is gravitational acceleration, Q is flow rate, and H_{net} is net head after deducting head losses from the gross head.

After connecting all the components with the energy sources data as inputs, the load, battery, and the converter, HOMER draws a schematic diagram of the proposed hybrid energy system as shown in Fig. 4.42. Wind generator, hydro turbine, and the diesel generator are connected to the AC electrical bus while the solar PV and the battery are connected to the DC electrical bus. A DC-AC-DC bidirectional converter is connected to both AC and DC buses.

4.14.3 Optimization results

After giving all the technical and economic input data to HOMER, we click on the calculate button, HOMER simulates the system and optimizes each possible configuration and presents the results both in tabular and graphical form. In tabular form HOMER presents two tables, sensitivity cases and the optimization results. In the sensitivity cases table, HOMER lists all the

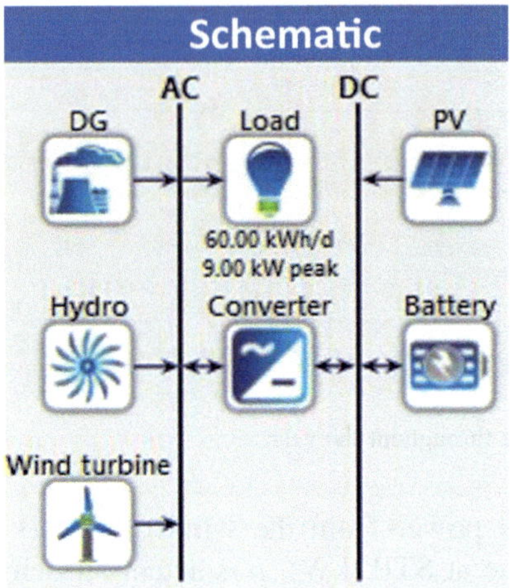

FIGURE 4.42 Schematic diagram of the proposed hybrid energy system.

Sensitivity		Architecture								Cost	
Wind Scaled Average (m/s)	Hydro Scaled Average (L/s)	PV (kW)	Wind Turbine	DG (kW)	Battery	Hydro (kW)	Converter (kW)	Dispatch		COE ($)	NPC ($)
4.00	10.0	3.73		10.0	16		4.00	CC		$0.370	$104,695
5.00	10.0	3.73		10.0	16		4.00	CC		$0.370	$104,695
6.00	10.0		1	10.0	8		6.00	CC		$0.361	$102,285
7.00	10.0		1	10.0	8		6.00	CC		$0.329	$93,258
4.00	15.0	3.73		10.0	16		4.00	CC		$0.370	$104,695

FIGURE 4.43 Sensitivity cases.

Export...

Optimization results

Left double click on a particular system to see its detailed simulation results.

Categorized Ov

Architecture							Cost			
PV (kW)	Wind Turbine	DG (kW)	Battery	Hydro (kW)	Converter (kW)	Dispatch	COE ($)	NPC ($)	Operating cost ($/yr)	Initial capital ($)
3.73		10.0	16		4.00	CC	$0.370	$104,695	$5,583	$32,519
		10.0	22		4.00	CC	$0.398	$112,765	$7,129	$20,600
2.19	1	10.0	7		6.00	CC	$0.444	$125,747	$5,654	$52,650
	1	10.0	10		8.00	CC	$0.452	$127,983	$6,249	$47,200
		10.0				CC	$0.525	$148,565	$11,028	$6,000
0.000814		10.0				CC	$0.525	$148,568	$11,028	$6,003

FIGURE 4.44 Optimization results.

possible cases within the search space ranked by the increasing NPC. In the sensitivity case table when we click on the one sensitive case, HOMER opens all the possible optimization results in the optimization results table for that particular sensitivity case.

Sensitivity cases of the proposed hybrid energy system are shown in Fig. 4.43. On clicking the first sensitivity case (highlighted in Fig. 4.43), optimization results of that sensitivity case are as shown in Fig. 4.44. Each row of the sensitivity case table and the optimization results table consists of the architecture of the system, cost of the system (NPC, cost

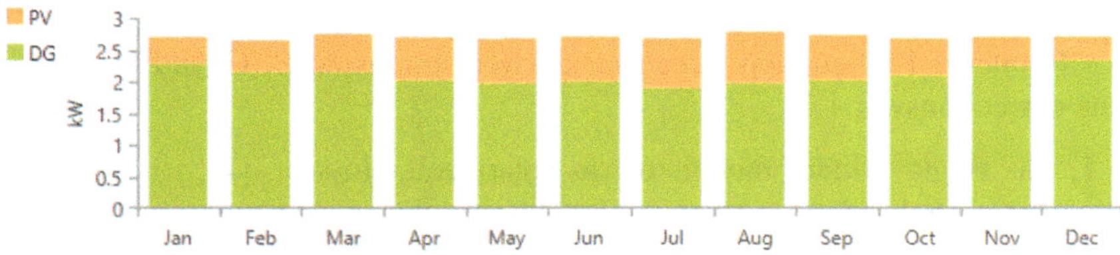

FIGURE 4.45 Monthly average electricity production.

TABLE 4.5 Cost summary of the first optimization case.

Components	Capital ($)	Replacement ($)	O&M ($)	Fuel ($)	Salvage ($)	Total
Converter	3600	1147	0	0	(646)	4100
Diesel	6000	14,636	7847	36,299	(811)	63,972
Li-ion	8000	11,687	2068	0	(535)	21,220
Photovoltaic	14,919	0	482	0	0	15,401
System	32,519	27,471	10,397	0	1993	104,694

of energy, operating cost and initial cost), system (renewable fraction, and total fuel), cost of each component, and the production of electricity from each source.

On a double click on any of the optimization results, HOMER opens the detailed simulation results of that particular optimization case. We discuss here the simulation results of the first optimization case of the proposed hybrid energy system. The first optimization result excluded the wind turbine and the hydro power from the hybrid energy system and the system produced electricity only from the solar PV and diesel generator. The simulation results page consists of the following tabs: cost summary, cash flow, compare economics, electrical, fuel summary, diesel, renewable penetration, Li-ion, PV, converter, and emissions. The share of electricity produced from solar PV and diesel generator is shown in Fig. 4.45. In a year, 5.174 kWh are produced from solar PV, whereas 18,367 kWh are from the diesel generator, summing up to 23,541 kWh/year. The consumption of electricity in a year is 21,900 kWh/year. One hundred and fifty five kilowatt hours are excess which could be exchanged to the grid if we use the net metering option provided in HOMER. The cost summary shows NPC of $104,695, $0.3698 as the COE, and $5583.11 as the operating cost. The detailed cost summary of the optimized case of the hybrid energy system is given in Table 4.5.

Problems

Problems 1–9 contain three/four answer options A, B, C, and D. Choose the correct answer.

1. In single diode model of a solar cell, losses due to mismatch between bandgap energies of solar irradiance and solar material are represented by
   - **A.** A diode
   - **B.** Series resistor
   - **C.** Shunt resistor
   - **D.** Current source
2. To increase the current from the solar panels, they are connected in
   - **A.** Series
   - **B.** Parallel
   - **C.** Series and parallel
   - **D.** All of above
3. By increasing the temperature of the solar cell, its open-circuit voltage
   - **A.** Increases
   - **B.** Decreases
   - **C.** Remains constant
4. By increasing the sun's irradiance, the short-circuit current of the solar cell is
   - **A.** Increased
   - **B.** Decreased
   - **C.** Remains constant
5. If the observation is $\Delta P < 0,\ and\ \Delta V > 0$, what will be the perturbation in the P&O algorithm?

   ΔD will be subtracted

   ΔD will be added

   - **A.** No perturbation
6. What will be the slope of the power curve at the MPP.

   $\frac{dP}{dV} > 0$

   $\frac{dP}{dV} < 0$

   $\frac{dP}{dV} = 0$
7. Which of the following is used in solar trackers?
   - **A.** Light-dependent resistor
   - **B.** Thermocouple
   - **C.** Thermistor
   - **D.** Light-emitting diode
8. Bandgap energy of the silicon is
   - **A.** 1.12 eV
   - **B.** 2.0 eV
   - **C.** 3.5 eV
   - **D.** 3.7 eV

9. Which of the following is the energy management software?
   A. HOMER PRO
   B. RETScreen
   C. iHOGA
   D. All of the above
10. Write down the five applications of solar thermal.
11. How does the increase in temperature deteriorate the performance of the solar cell?
12. Write down the five software names that are used for energy management.
13. Differentiate between the solar tracker and the MPPT.
14. Write down the working principle of a solar water heater.
15. At STC, if the $I_{Ph} = 2.5A$, and $I_0 = 0.2A$, what would be the open-circuit voltage (V_{oc}). If FF is 0.54 and the input power is 10 W, what would be the efficiency of the solar cell.
16. What is the solar photovoltaic grid parity?

References

[1] S.C. Bhatia, 4—Solar thermal energy, in: S.C.B.T.-A.R.E.S. Bhatia (Ed.), Advanced Renewable Energy Systems, Woodhead Publishing India, 2014, pp. 94–143.

[2] Z. Salameh, Chapter 5—Emerging renewable energy sources, in: Z.B.T.-R.E.S.D. Salameh (Ed.), Renewable Energy System Design, Academic Press, Boston, 2014, pp. 299–371.

[3] A. Herez, M. Ramadan, M. Khaled, Review on solar cooker systems: economic and environmental study for different Lebanese scenarios, Renew. Sustain. Energ. Rev. 81 (2018) 421–432. Available from: https://doi.org/10.1016/j.rser.2017.08.021.

[4] M. Kamran, M. Bilal, Z.J. Zaib, LabVIEW based simulator for solar cell characteristics and MPPT under varying atmospheric conditions, Mehran Univ. Res. J. Eng. Technol. 37 (3) (2018) 529–538. Available from: https://doi.org/10.22581/muet1982.1803.07.

[5] M. Kamran, M. Mudassar, M. Rayyan, M. Usman, M. Bilal, R. Asghar, Implementation of improved perturb & observe MPPT technique with confined search space for standalone photovoltaic system, J. King Saud. Univ.—Eng. Sci. 32 (7) (2020) 432–441. Available from: https://doi.org/10.1016/j.jksues.2018.04.006.

[6] H. Shahid, M. Kamran, Z. Mehmood, M.Y. Saleem, M. Mudassar, K. Haider, Implementation of the novel temperature controller and incremental conductance MPPT algorithm for indoor photovoltaic system, Sol. Energy (2018). Available from: https://doi.org/10.1016/j.solener.2018.02.018.

[7] S. Qazi, Chapter 7—Solar thermal electricity and solar insolation S. B. T.-S. P. (PV) S. for D. R. and, in: R.A. Qazi (Ed.), Standalone Photovoltaic (PV) Systems for Disaster Relief and Remote Areas, Elsevier, 2017, pp. 203–237.

[8] Z. Liu, Chapter 6—Innovation in global energy interconnection technologies, in: Z.B.T.-G. E.I. Liu (Ed.), Global Energy Interconnection, Academic Press, Boston, 2015, pp. 239–272.

[9] M.U.H. Joardder, P.K. Halder, M.A. Rahim, M.H. Masud, Chapter eight—Solar pyrolysis: converting waste into asset using solar energy M.G. Rasul, A. Kalam Azad, and S. C. B. T.-C. E. for, in: S.D. Sharma (Ed.), Clean Energy for Sustainable Development, Academic Press, 2017, pp. 213–235.

[10] T. Funabashi, T.B.D.E.R. Chapter 1—Introduction T.-I. ofin, in: P.S. Funabashi (Ed.), Integration of Distributed Energy Resources in Power Systems, Academic Press, 2016, pp. 1–14.

[11] P. Breeze, Chapter 13—Solar power P. B. T.-P. G. T. (Third E. Breeze), Solar Power Generation, Newnes, 2019, pp. 293–321.

Chapter 5

Wind energy

Muhammad Rayyan Fazal and Muhammad Kamran
Department of Electrical Engineering and Technology, Riphah International University, Pakistan

5.1 Introduction

The wind is moving atmospheric air that is clean, inexhaustible, ubiquitous, and free at the same time. It is generated due to three main reasons: the revolving Earth, irregular patterns of the Earth's surface, and uneven heating from the sun in the atmosphere. The wind flow patterns are shaped through buildings, environmental factors, and Earth's terrain. This wind can generate electricity if harvested by employing wind turbines. Windmills are used in various applications but those generating electricity are termed wind turbines. The kinetic energy of wind is converted to a usable form of energy through these turbines. This kinetic energy of wind causes the turbine blades to rotate. A shaft coupled with the electrical generator is attached to these blades. The generator then converts the mechanical power to electrical power.

The wind flow was used in the Nile for the first time around 5000 years ago to power the boats. It was later on used by Europeans for water pumping and grinding grains in the 1700s and 1800s. Professor James Blyth, in 1887, invented the first wind turbine in Scotland, Glasgow at Anderson's College [1]. In 1931 Russians installed a 100-kW horizontal axis wind turbine (HAWT) connected to the local 6.3 kV system. The world's first-megawatt turbine was installed in Castleton, Vermont, US in 1941. The invention of the two-bladed wind turbine in Denmark replaced the DC-type electrical machines with AC machines as wind machines in the 1950s. Three-bladed wind turbines were invented at the same time by a research group in Denmark. The oil crisis of 1973 led countries like the United States, United Kingdom, Sweden, and Germany to invest in wind energy. The early years of the 1980s were considered to be revolutionary for wind turbine technology with the invention of the Tvind generator of 2 MW power. It was the first time in history that wind power was controlled using power electronics [2]. During the same duration, Nordtank wind turbine generators of 55 KW capacity caused the per kWh cost of wind energy to drop by half. Thousands

Renewable Energy Conversion Systems. DOI: https://doi.org/10.1016/B978-0-12-823538-6.00003-8

of these generators were sent to wind farms in California by that time [3]. By 2016 wind power was contributing 4% of the total power worldwide. It has been estimated that by 2030, 20% of the total electricity of the world will be generated through wind turbines [4]. The world's capacity for wind energy reached 650.8 GW in 2019, the second-best year for market-size with a 10.1% growth rate [5]. However, wind energy is expected to be the most prominent source of electricity by 2050 with an estimate of providing 35%, more than one-third of the world's need, of electricity [6].

Since then, the technology has advanced day by day becoming more efficient, and more reliable with higher output power, that is, from a few kilowatts (kW) to several megawatts (MW). In 2020 its many advantages are the driving force behind the rapid advancement of wind energy technology:

- The wind is free and plentiful, nations can get a great advantage out of it as unlike other fuel sources (fossil fuels), it is independent of uncontrollable price hikes. Hence it is free from the uncertainty of price that fuel costs add to the conventional energy sources.
- Over the past few years, the price associated with wind turbines has decreased sharply due to improved technology, higher efficiency, lower prices of associated power electronics, and mass production. The associated cost of electricity depends mainly upon project financing and available wind resources. However, wind farm electricity can deliver power at a fixed price for more than 20 years. The investment for a wind turbine is estimated to be retrieved within 7 years with a life expectancy of over 30 years.
- The pollution-free nature of wind energy is another compelling reason for its evolution. Oil and gas pumping and mining of coal and uranium have serious environmental impacts on sea and land conditions. Also, the combustion of fossil fuels pollutes the air, emitting particulate matter, nitrogen oxide, and sulfur dioxide—causing serious health issues (asthma, Alzheimer's disease), smog, acid rain, climate change, and economic constraints. Whereas wind farms are benign in this regard and provide electricity without any emissions; a very small amount of GHG emissions are produced during the manufacturing and transportation process of turbines and blades. A 1 MW wind turbine is estimated to counterbalance 2360 tons of CO_2 in the atmosphere.
- Wind turbines can be installed at any place with good wind potential. They can be erected near landfill sites, existing farms, pasturelands, main roads, sides of motorways, and any other place where urban development is unlikely. Furthermore, they reduce the cost of electricity transportation over great distances as the cost of cables, towers, and electricity losses are eliminated.
- The wind is, in fact, another form of solar energy, hence it is sustainable. Sunshine causes winds to blow, which can be harnessed to send energy to the power grid making it a good choice for investment. In the future,

stronger winds are expected to be produced because of global warming and changing climatic conditions.

- Wind farms are commissioned in a relatively short time, that is, 2–3 years, as compared to nuclear power plants that require decades to be designed and built.
- Another prominent advantage of wind turbines is the conservation of water resources. The Civil Society Institute of the United States published a report in September 2012 that concluded that large volumes of water are used by conventional power stations (oil, gas, coal, and nuclear). The solar-powered setup uses 225–520 gal/MWh water for washing PV panels, nuclear uses 700–100 gal/MWh, biomass uses 40,000–100,000 gal/MWh, and coal uses 500–600 gal/MWh of water. Whereas wind, as compared to others, requires a negligible amount of water for its operation. But other phases of the wind turbine lifecycle involve water usage, which is estimated to be 45–85 gal/MWh [7].
- The wind turbine industry has great potential for job creation in manufacturing, commissioning, transportation, maintenance, and providing support services.
- Wind turbine projects can provide economic benefits in rural areas. They can earn a good profit for landowners as land is leased to electricity companies for 25–50 years. Farmers and ranchers can continue routine work, the only fraction of land is used with no interference from wind turbines providing landowners additional income.

There are many challenges for the wind turbine industry associated with capturing wind power.

- Despite a sharp decrease in operational cost, still, an initial investment in wind turbines is huge as compared to fossil fuel-based generators. However, governments around the globe are providing subsidies due to lifetime environmental and economic benefits.
- There is uncertainty related to wind, as energy may not be required when the wind starts blowing or energy may be required when the wind stops. This problem can be overcome by energy storage, using different means, when the wind is available and consuming it when required.
- The best windy locations are in hilly areas, far away from the urban locations. Hence to bring wind power from remote locations into cities, a transmission line network is required.
- Wind turbines have less impact on the environment when compared with other electricity plants, however, they are a bit noisy due to the moving turbine blades. Aerodynamic and mechanical noises are produced during wind turbine operation. Aerodynamic noise is the dominant source of noise produced because of the howling sound caused by the airflow through turbine blades. Aerodynamic noise can be reduced by designing the turbine blades carefully. Mechanical noise is produced because of the

various components like hydraulic system, gearbox, and generator. This noise can be minimized using sound insulation inside the turbine, soundproof curtain, and antivibration support [8]. This noise decreases exponentially as the distance from the tower increases. The wind turbine located 350 m away from home is not noisier than a home refrigerator. Noise remains just 35 dB if the distance of 500 m is maintained from the wind turbine which is even less than a normal conversation that is rated to be 60 dB [9].

- Millions of deaths of different birds have been reported per year due to wind turbines. However, this problem is expected to be resolved through technological advancements. There are fewer deaths reported due to modern slow turning blades as compared to older fast turning blades. Also, the wind industry is trying to ensure that migrating birds would not get attracted to wind turbine lights during foggy nights and bad weather conditions. Minimal lights have to be installed and the glare of these lights must be prevented using advanced techniques [10].
- There are some concerns related to the visual beauty of the natural landscape.
- Disco effect or shadow flicker is the reflection of sunlight caused by the turbine body or moving blades. It changes with the intensity of light and causes disturbance for the nearby living residents. This effect can be minimized by coating the body and blades of the turbine with less reflective material [11].
- A major safety concern related to wind turbines is the possibility of blades being damaged or falling apart. It can seriously damage the tower or any other nearby living beings. Hence, it is suggested to install the wind turbines away from residential areas.
- The latest wind turbines require strong permanent magnets to generate electricity at slow speeds. There is a shortage of the rare earth element neodymium (Nd) that is required to make such magnets. China was leading the world in producing rare earth metals but now mines have been reopened in the United States to contribute to meeting the increasing demand.
- Wind turbine technology is new and there is a shortage of technical staff. it becomes problematic if any wind turbine starts behaving abnormally, especially in rural areas. However, maintenance requirements for wind turbines are much less than for most of the other electricity generating equipment.

The future of wind energy is very bright, this is mainly because of the decline in fossil fuel reserves, the huge costs associated with fossil fuel exploration, and the shutdown of coal power plants due to environmental hazards. Developing countries can take the maximum benefit out of it, especially where energy poverty is a big challenge. The wind turbines can be economically viable in rural areas and remotely located areas with no access to grid electricity. It is a very fast-growing market that is getting mature. Developing countries invested heavily in the wind to cater to environmental problems but now it is economically well justified too. Well-developed and industrialized countries are now experimenting with renewables on a big scale, as an example, in 2016, 95% of German's Electricity was delivered using renewable energy resources (36% wind energy) on May 11, at 08:00 a. m. This scenario made power prices go so negative that the German Government paid the commercial customers to use electricity for that particular time [12]. Norway is providing 100% wind electricity to one of its islands with a population of around 4000. Twenty-one turbines are supplying 30×10^6 kWh to the local population and 80×10^6 kWh to the national grid. Denmark is leading the European market in turbine manufacturing as they took the initiative in 1985 to invest in renewables. They are getting 39% of their total electricity from wind only, and the trend is likely to continue [13]. Now the next challenge is to find the optimum solution for storing electricity when the wind blows but electricity is not required. However, it will remain a challenge to continuously look for more efficient and economical ways of energy storage.

This chapter covers all aspects of wind turbine technology, allowing students to understand basic to advanced concepts

5.2 Wind energy fundamentals

The wind is the motion of atmospheric air traveling from equatorial regions toward the Earth's poles. The wind might feel peaceful, invisible, and delicate or loud, destructive, and visible, depending upon the speed of air. The sun heats the Earth's surface unevenly, stronger in tropical regions, and weaker in higher altitudes. Also, water, soil, and rock present at various locations have a different warm-up rate. "Convection" converts this unbalanced nature of heating to air motion which is further controlled by the motion of the Earth. Convection is regarded as the transfer of heat from one place to another due to molecular movement. Higher specific densities of landmasses allow them to transfer heat better than the water bodies. These lands are therefore heated faster by the sun during daylight causing air to rise, creating lower inland pressure, and attracting cooler air from the sea. The pressure of air at night is higher at ground level causing air to blow in the reverse direction.

The warm air becomes lighter and rises upwards, whereas cool air gets heavier and descends onto the surface of the Earth. Hence cool air travels from the north pole to the equator and gets warm on its way, whereas warm air rises to get cool and heavy until it starts descending back toward the poles. There exists a phenomenon named the "Coriolis effect," diverting the northern hemisphere winds to the right and the southern hemisphere winds to the left. The Coriolis effect causes the wind's direction to change which is referred to as Ferrel's law. It states that any horizontally moving object or wind moving freely will be deflected to the right in the northern hemisphere and Earth's rotation will deflect them to the left in the southern hemisphere. Some regions are having lower or higher atmospheric pressure than the average causing the winds to blow from a higher toward a lower pressure region. This relocating wind in a cyclone experiences inward as well as outward force. Inward force is regarded as strain gradient and outward forces include centrifugal and Coriolis. For a cyclone, the relationship between inward and outward forces can be given as

$$\text{Strain Gradient Force (Inward)} = \text{Coriolis Force (Outward)} + \text{Centrifugal Force (Outward)}$$

The outward forces balance the inward force pressure, wind running in this fashion is regarded as "gradient wind flow." The overall wind flowing mechanism is presented in Fig. 5.1.

These phenomena affect the global winds at large. Therefore the rise of winds at the equator causes pressure to drop near ground levels attracting northern and southern winds. The cool air present at the poles causes the high pressure of winds. Hence the study of the geographical data is of extreme importance to get the most appropriate site for wind turbines. The

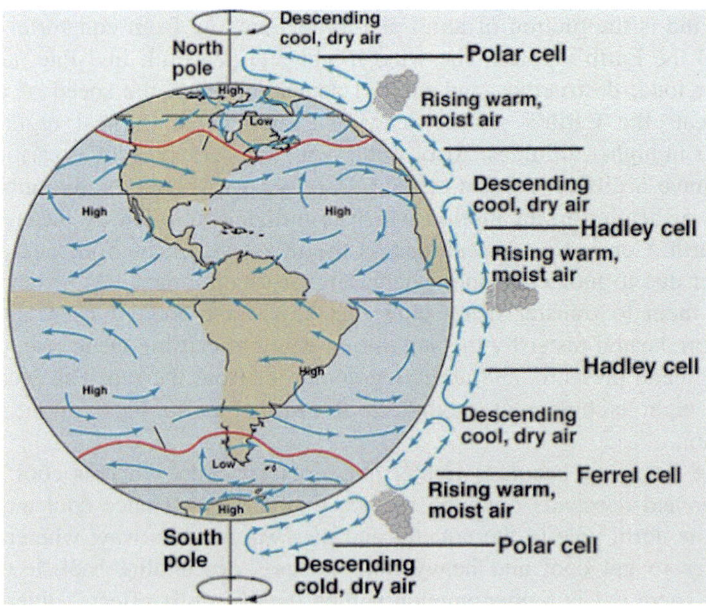

FIGURE 5.1 Wind flowing mechanism [14].

local topology of the land also affects the direction, speed, and overall power of the winds.

Structural and mechanical designing of wind turbines are required according to the various parameters and conditions, including static loading, dynamic loading, fatigue loading, the minimum amount of wind required to start the equipment, transportation, commissioning, installation, and flexibility of loads. To capture the maximum wind power, a larger diameter of blades must be used. But it is not that easy to install large blades and big rotors as there must be a balance between the weight and structure of the wind turbine. Also, it depends upon the surrounding conditions. The rotor design is very significant as the rotor and its blades directly affect the efficiency of the turbine, soft designing techniques are used for this purpose. Turbine blades are carefully designed to reduce the weight and increase the efficiency of the wind turbine. Towers' design is of extreme importance as they have to bear overall structure along with wind and climate variations. Fatigue cracks due to excessive vibrations are a major reason for the turbine towers to collapse, but there are very few cases of it. However, it did happen in Scotland when two towers collapsed in early 2008, the associated loss was expected to be in millions of dollars [15]. Highways must also be in good condition to handle the transportation requirements of wind turbines.

The wind turbine design is associated with the following characteristic parameters.

- *Start-up speed*: it is the minimum speed of the wind at which turbine blades start running after being at rest position.
- *Cut-in speed*: it is the speed of the wind at which the generator starts producing power and a trickle of electricity is produced as a result. At cut-in speed, the turbine generator can be used as a motor to overcome the inertia required to turn the blades. Typical cut-in wind speed is in the range of 7–9 mph.
- *Rated speed*: it is the minimum speed of the wind at which the generator delivers advertised power. Blades regulation is activated at this speed. Typically rated wind speed is in the range of 25–35 mph. However, small and medium-sized wind turbines produce more power than their rated speed.
- *Furling speed*: it is the maximum speed of wind at which the turbine can develop power beyond which the machine is forced to shut down to avoid damage. At this wind speed, turbine blades start folding toward the tail using tail vane. Hence rotor area facing the wind is reduced, protecting them against high winds.
- *Cut-out speed*: it is the wind speed at which the protection mechanism of the turbine activates to stop the generator from producing power. Different techniques are in practice to physically stop the rotor (brakes) and blade feathering (changing blade angles) against high winds. The typical cut-out wind speed is nearly 55 mph.

- *Cut-back-in speed:* this is the speed at which wind has to return for a certain period to restart the action of the Wind turbine. The typical cut-back-in wind speed is nearly 45 mph.
- *Rated power*: this is the advertised power for which a wind turbine is designed to produce under specific conditions.
- *Peak power*: it is the maximum amount of power a generator of wind turbines can deliver.

It is, in fact, important to note that wind speed variation is a continuous phenomenon and therefore windows are created to describe the particular wind speed. for instance, 5 mph means wind speed is between 4.5 and 5.5 mph, 5 m/s means wind speed is between 4.75 and 5.25 m/s.

5.2.1 Types of winds: meteorology

The science of Meteorology involves the atmospheric study of all the wind phenomena to observe and predict different weather processes. It is a branch of atmospheric sciences involving atmospheric physics and atmospheric chemistry. It helps to understand the theoretical and practical perspectives of the wind phenomenon. We can classify the wind into three main types that exist on our planet.

1. Planetary winds
   a. Trade winds or tropical easterlies
   b. Antitrade or westerlies
2. Periodic winds
   a. Land and sea breeze
   b. Mountain and valley breeze
   c. Monsoon winds
3. Local winds

Many mathematical modeling tools and algorithms are used to simulate the wind behaviors that are utilized to analyze, evaluate, and forecast. Groups of weather scales are created to categorize different weather phenomena according to the life span of wind systems and its horizontal span. These scales from largest to smallest can be written as [16]:

- Planetary scale: planetary scale weather, also called global scale weather, has the largest span, around 10,000 km with the longest life span. They cover the globe from one end to another and may last for several weeks. Examples include:
  - Polar jets
  - Jet stream
  - Trade winds
  - Antitrade or westerlies
- Synoptic-scale or large-scale weather: Synoptic is a Greek word that means “seen together” also called overall view. Synoptic meteorology is

a study of disturbances caused by weather on a large-scale over large areas to have a clear understanding of the current atmospheric state. Weather maps are used in synoptic meteorology for the large-scale weather forecast. Weather maps and local weather forecasts are nothing but synoptic-scale meteorology. These systems span larger distances varying from a few hundreds to thousands of kilometers, typically 1000 km with life spans of a few days to some weeks. Examples include:
  - Air masses
  - Mid-latitude cyclones
  - Tropical cyclones
  - Extratropical cyclones
  - High- and low-pressure systems
  - Hurricanes
  - Typhoons
- Mesoscale weather: these winds are produced due to orographic lift and thermally direct and indirect air circulations. Weather conditions that cannot be shown on weather maps as they are so small are regarded as mesoscale. These events can impact regional or local levels ranging from 10 to 100 km with a life span of 24 hours or even less. Examples include:
  - Sea breeze
  - Tornadoes
  - Thunderstorms
  - Land breeze
- Mesoscale meteorology is the study of the impact of regional topography on climatic conditions to give rise to mesoscale events. There are some events even smaller than mesoscale events referred to as microscale events. They exist for a very short period with a span of 100–1000 m. their contribution is negligible in trimming our daily weather conditions. Examples include:
  - Sand devil or dust devil
  - Turbulence

5.2.2 Capturing the wind: wind speed, energy, and power

Wind turbines work on the principle of capturing the kinetic energy of the wind to transform it into rotational mechanical energy with the help of a rotor. The shaft in the wind turbine rotates due to the rotational movement of the rotor. This movement of the shaft can be then directly used to drive a generator or for heat production or for pumping water. Hence wind can be converted into heat energy, electrical energy, or mechanical energy using appropriate equipment. Power is a term used most commonly to describe the performance of any machine, it can be described as the energy extraction per unit of time. Wind power is dependent on several key factors:

- Air density (mass per unit volume)
- Air speed

- Area of the rotor
- Area of the blades

Wind power indicates the amount of energy that a blowing wind can produce in a particular area per unit of time. Whereas wind energy can be expressed as the amount of energy present in the air due to its movement. Winds create a torque that causes turbine blades to rotate. Winds become stronger at high densities and their kinetic energy depends upon the air density. Therefore the denser the air is, the more the kinetic energy will be produced [3]. Whereas air density depends upon the temperature and pressure of the air. It decreases with an increase in altitude or temperature or humidity, whereas it is directly proportional to the air pressure. Colder winds are denser than hot ones, hence they carry a higher amount of kinetic energy. At 15°C under normal atmospheric pressure at sea level, that is, 1 atm (14.7 psi), air density can be given as 1.225 kg/m^3 which decreases with an increase in temperature. Air density of the dry air can be calculated as [17]:

$$\rho = \frac{P}{RT} \tag{5.1}$$

ρ is air density (kg/m^3)P is pressure (pa)R is specific gas constant (287.058 J/kg · K for dry air)T is temperature (K).

Temperature and pressure are varying quantities that vary with the change in altitude. A single equation of air density calculation considering the combined effect can be written as:

$$\rho = \rho_o e^{-\left\{\frac{0.297H_s}{3048}\right\}}$$

This equation remains valid for up to 6 km (20,000 ft) above sea level where H_s (m) states site elevation. The simple form of this equation is:

$$\rho = \rho_o - \left(1.194 \times 10^{-4} H_s\right)$$

The higher attitude will significantly affect the air density. For example, at 2500 m (8200 ft), the air density reduces to 0.9265 kg/m^3, that is, 25% lower than the standard value of sea level.

The variation of temperature (Celcius) along with attitude can be found as:

$$T = 15.5 - \frac{19.83H_s}{3048}$$

Blades area (diameter) is another factor that affects wind energy. Larger blades mean larger diameter or more rotor area, hence more wind can be captured under the same atmospheric conditions. Kinetic energy of the wind can be expressed as:

$$KE_w = \frac{1}{2}mv^2 \tag{5.2}$$

(whereas associated terms can be given as $m = \rho V$, $V = Ad, A = \pi l^2$)

Eq. (5.2) becomes

$$KE_w = \frac{1}{2}\rho\pi l^2 dv^2 \tag{5.3}$$

Overall wind power can be defined as the rate of flow of kinetic energy, it can be given as:

$$P_t = \frac{KE_w}{t} \tag{5.4}$$

Using Eqs. (5.3, 5.4) becomes

$$\begin{aligned} P_t &= \frac{\frac{1}{2}\rho\pi l^2 dv^2}{t} \\ &= \frac{\rho\pi l^2 dv^2}{2t} \\ &= \frac{1}{2}\rho\pi l^2 v^2 \frac{d}{t} \end{aligned} \tag{5.5}$$

Distance per unit time is velocity, that is, $v = d/t$

The final form of the power equation is

$$P_t = \frac{1}{2}\rho\pi l^2 v^3 = \frac{1}{2}\rho A v^3 \tag{5.6}$$

where P_t is wind power (W), ρ is density (kg/m^3), m is air mass (kg), v is air velocity (m/second), V is volume (m^3), A is area covered by the blades (m^2), d is the distance traveled by air (thickness) (m), and l is length of the blade (m)

This is a fundamental equation for expressing the wind power that highlights a linear relationship with the air density and a nonlinear relationship with the velocity of air and length of blades. Air density fluctuations have limited range, making them of less significance as compared to wind speed and length of blades. Power increases with the square of the blade's length and the cube of the wind speed. It underlines the importance of both these factors. A threefold increase in the wind speed will lead to a 27-fold increase in the output power. This high dependency on wind speed forces us to adopt the method so that the speed of the wind can be accurately measured. This is accomplished with the help of anemometers. An anemometer is an instrument used to measure the wind speed, some of them are capable of measuring wind direction too. The cup anemometer is the most commonly used consisting of three or four cups. These cups are mounted so that wind forces them to rotate at a speed proportional to the speed of the wind. The number of revolutions per unit time are recorded using data acquisition systems; this data can be stored for 1 year or more [18]. A windmill anemometer along with a wind vane is used to measure wind speed and direction, respectively. In the latest anemometers, speed of sound is used for the precise wind speed measurements Table 5.1.

TABLE 5.1 Density-based classification of wind power [19].

Potential	Wind class	10 m (33 ft) height		50 m (164 ft) height		80 m (263 ft) height	
		Wind speed (m/second)	Wind power density (W/m^2)	Wind speed (m/second)	Wind power density (W/m^2)	Wind speed (m/second)	Wind power density (W/m^2)
Fair	Class 1	0–4.4	0–100	0–5.5	0–200	0–5.9	0–250
	Class 2	4.4–5.1	100–150	5.5–6.4	200–300	5.9–6.9	250–380
Good	Class 3	5.1–5.6	150–200	6.7–7.0	300–400	6.9–7.5	380–500
	Class 4	5.6–6.0	200–250	7.0–7.5	400–500	7.5–8.0	500–600
Excellent	Class 5	6.0–6.4	250–300	7.5–8.0	500–600	8.0–8.6	600–750
	Class 6	6.4–7.0	300–400	8.0–8.8	600–800	8.6–9.4	750–980
Exceptional	Class 7	7.0–9.5	400–1000	8.8–12	800–2000	9.4–12.8	980–2400

There are seven classes of wind power [4]. These classes are meant to provide information about the specific location as these classes are included in wind maps. These maps provide information regarding wind speed that is vital in wind turbine siting.

National Renewable Energy Laboratory provides a standard wind resource classification using wind power density. Wind power density compares various wind resources being independent of the size of the wind turbine. As compared to the wind speed, it provides better insight into the comparison of site alternatives. The main reason involves the nonlinear cubic dependence of wind power on speed. Various classes of wind and their corresponding densities at wind speed at different heights are presented in the table.

5.3 Potential and prediction of wind energy

5.3.1 Wind assessment

Wind turbines can operate independently from distant locations or connected to the grid. The most important aspect for the site selection of wind turbines is the wind resource measurement and assessment through initializing a solid program. It considers the speed and quality of wind as the most critical factors. The cubic relationship with power makes the wind speed one of the most crucial factors for the wind turbine siting. It is suggested to take at least 10-year data for the mean wind speed, however, the period is too long for a project to handle and expensive too. So, short-term measurement, that is, 1 year is done through the measure, correlate, and predict scheme. It is essential to formulate a quality plan and associated structure for every wind program. The selection of apparatus, equipment, and associated measurements having adequate standards, reliable data collection, and analysis schemes are the most significant aspects to take care of.

The other two major factors are terrain roughness and friction coefficient. Friction coefficients relate to the characteristics of the terrain and various roughness length patterns are given in the table. Friction increases with an increase in height of anything present on the ground. The relationship of wind speed, friction coefficient, and terrain height is given as:

$$\left(\frac{s}{s_o}\right) = \left(\frac{h}{h_o}\right)^{\alpha} \tag{5.7}$$

s_o represents the known wind speed associated with h_o height of the terrain, whereas s is wind speed related to the h height of the terrain, α is the friction coefficient with the values between 0 and 0.4 [20]. This equation is used to calculate the reduced rate of wind speed. It can also be calculated by considering roughness length l:

$$\left(\frac{s}{s_o}\right) = \frac{ln(h/l)}{ln(h_o/l)} \tag{5.8}$$

There is an exponential relationship between wind speed and altitude. Hence, wind speed at 100 m will be different from the 200 m height. The relevant formula to calculate the wind speed at any height is:

$$s = s_k \frac{ln(l/l_o)}{ln(l_k/l_o)} \tag{5.9}$$

Here s is the speed at height l, s_k is the known wind speed at a height l_k, l_o is the length of roughness in direction of the wind.

For wind turbine siting, class 0 indicates the ideal location that includes the constant wind, steady speed, and absence of any turbulence. Class 0 refers to the open lands and water surfaces where ideal conditions exist because low turbulence also increases the lifecycle of the wind turbines. Hence to get the maximum out of it, many of the wind turbines are sited offshore, that is, 5–10 km into the sea. Land obstacles such as buildings, mountains, forests are unable to affect the performance of wind turbines located offshore. Therefore based upon their siting, there are two types of wind turbines, onshore wind turbines (situated on land) and offshore wind turbines (situated in the sea). Power transmission from offshore turbines is however difficult compared to onshore turbines.

5.3.2 Turbine power assessment

5.3.2.1 Betz law

Betz law provides the maximum theoretical power of the wind. The wind carries kinetic energy that is changed once it cuts the turbine blades. The speed of the wind approaching a wind turbine is significantly reduced after it passes through it. Therefore there are two types of wind speeds, that is, higher speed before reaching the wind turbine and lower speed after it passes through the wind turbine. Wind turbines are specifically designed to pull out this kinetic energy of the wind. The amount of the extracted power can be assessed by analyzing the reduced wind speed (after it passes through the turbine) [21]. The kinetic wind energy can be given as:

$$KE_w = \frac{1}{2}\rho\pi l^2 d v^2 \tag{5.10}$$

If wind speed after and before passing through wind turbine is v_1 and v_2 respectively then the kinetic energy will adopt the form as:

$$KE_w = \frac{1}{2}\rho\pi l^2 d\left(v_2^2 - v_1^2\right) \tag{5.11}$$

Whereas the maximum available extracted power P_E can be given as:

$$P_E = \frac{KE_W}{t} = \frac{d}{2t}\rho\pi l^2\left(v_2^2 - v_1^2\right) \tag{5.12}$$

Whereas πl^2 is the area and d/t is the wind speed, which in this case becomes $\frac{v_1+v_2}{2}$, hence giving the values,

$$P_E = \frac{KE_W}{t} = \frac{1}{2}\rho A\left(\frac{v_1+v_2}{2}\right)\left(v_2^2 - v_1^2\right) \tag{5.13}$$

The relation of total and extracted wind power can be represented by coefficient C as follows:

$$C = \frac{P_E}{P_t} = \frac{\frac{1}{2}\rho A\left(\frac{v_1+v_2}{2}\right)\left(v_2^2 - v_1^2\right)}{\frac{1}{2}\rho A {v_2}^3} = \frac{1}{2}\frac{(v_1+v_2)\left(v_2^2 - v_1^2\right)}{{v_2}^3} \tag{5.14}$$

As we have defined the wind speed v_2 as speed before passing through the wind turbine, so the wind speed of total power, that is, v can be here replaced by v_2.

It can be simplified as:

$$C = \frac{1}{2}\frac{(v_1+v_2)\left(v_2^2 - v_1^2\right)}{{v_2}^3} = \frac{1}{2}\left(1 - \frac{v_1^2}{v_2^2}\right)\left(1 - \frac{v_1}{v_2}\right) \tag{5.15}$$

For maximum available power, we can take differential of this coefficient w.r.t. v_1/v_2 to solve it for v_1/v_2

$$\frac{d(C)}{d(v_1/v_2)} = 0,$$

$$\frac{d(C)}{d(v_1/v_2)} = \frac{1}{2}\left[-3\left(\frac{v_1}{v_2}\right)^2 - 2\left(\frac{v_1}{v_2}\right) + 1\right] \tag{5.16}$$

Combining both

$$\frac{1}{2}\left[-3\left(\frac{v_1}{v_2}\right)^2 - 2\left(\frac{v_1}{v_2}\right) + 1\right] = 0 \tag{5.17}$$

Hence it follows:

$$3\left(\frac{v_1}{v_2}\right)^2 + 2\left(\frac{v_1}{v_2}\right) - 1 = 0 \tag{5.18}$$

Solving yields $v_1/v_2 = 1/3$

Hence putting back in equation C gives the maximum value as

$$C \approx 59\%$$

This power coefficient provides an insight into the maximum available power against wind, that is, two-thirds of the total available wind speed. It indicates that it is impossible to extract all the wind energy because after rotating the wind blades, wind speed cannot go to zero.

5.3.3 Estimating wind power

Wind power estimation is directly linked to the wind turbine site selection for any particular site. A successful wind resource assessment program is crucial for proper site selection to look for the expected wind energy in that area. Hence a proper framework and solid quality insurance plan must be developed through applying standard schemes of data collection, analysis, and assessment. This approach will help estimate wind energy production over a certain period. The wind resource assessments can be performed at micro-, meso-, and macroscales. Microscale assessment is confined to the small area or local level estimation, mesoscale is linked to the national level estimation and macroscale study is required for global estimation. Wind maps are useful tools for wind assessment but some extra effort is required for proper judgment. These additional points can be presented as [22]:

1. Land availability
2. Terrain characteristics
3. Instantaneous measurement of wind speed
4. Feedback from stakeholders regarding the environmental impact of wind turbines
5. Meteorological data analysis related to speed and direction of the wind

There is extensive literature available for wind assessment studies, and Asian countries are believed to be a future hotspot in terms of investments in wind energy due to enormous untapped potential. The proper selection of the site must also include the modeling associated factors like proper software selection, wind distribution, overall and individual ratings of components, and overall energy potential [23–28].

5.3.4 Predicting wind energy

The intermittent nature of wind makes it necessary to look for a possible way to forecast wind energy. This forecasting is different from wind energy estimation in the sense that the former relates to the wind speed prediction at any time. Whereas energy estimation provides the potential of wind energy related to that area. Wind energy-related scheduling and energy dispatch is improper so far, hence the prediction can help the overall system in several ways, including efficient scheduling for wind energy accommodation, better economic constraints, and allows the stakeholders to look for alternate options to avoid penalties. Many sophisticated high-accuracy models have been developed for wind speed prediction [29–33]. Many are already in use by different wind energy producers.

5.4 Wind energy conversion systems

5.4.1 Basic components of wind turbine

Wind turbines mainly comprise a tower, nacelle, turbine, controlling, and sensing mechanism. The basic assembly of wind turbine systems is shown in Fig. 5.2. Wind turbines are designed so that their speed can be controlled or even stopped against high wind speed or maintenance purposes. Modern systems have their controlling mechanism supported by various components. They have anemometers, different sensors for electrical and mechanical system monitoring, a wind sensing mechanism to move the turbine according to wind direction, stall controllers responsible for starting and shutting down the machine against threshold values of wind speed, power, and control electronics for conversion and controlling, battery for storage, and transmission link to the grid.

5.4.1.1 Turbine

"Low-speed rotor" or turbines contain two to six blades, with three-blade mechanisms most common. Three-blade turbines are more stable, efficient, and lightweight. Whereas two-blade turbines are inefficient in capturing wind and require a greater amount of wind speed to turn on the turbine at the beginning. The turbine sizing ranges from a few kW, mostly used for standalone remote areas, to a few MW for large-scale generation. The turbine blades are aerodynamic and designed to be as light as possible. They are made up of composite materials like high-density wood, carbon, or

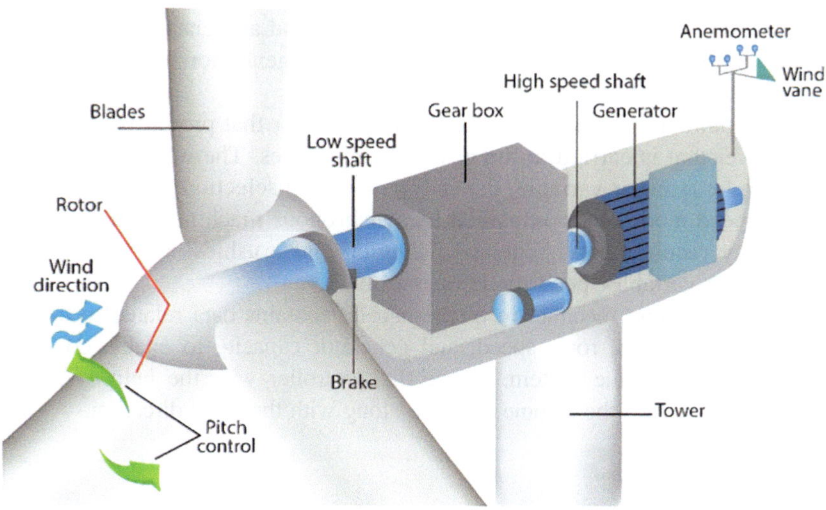

FIGURE 5.2 Various wind turbine components [34].

glass-fiber based Plexiglas composites. The increased radius of turbine blades can capture a greater amount of wind. The turbine power is kept constant by employing the "pitch control" mechanism which is activated at higher wind speeds by changing the effective blade surface area. "Pitch control" can be defined as turning or pitching out of the wind to control the amount of wind for wind speeds below or higher than a certain threshold.

5.4.1.2 Yaw control system

The yaw control mechanism comprises a motor and drive. The main purpose of this arrangement is to move the nacelle and blades according to the wind direction. It enables the wind turbine to capture the maximum available wind. During the nacelle movement, a fair chance of cable twisting occurs inside the tower. The cable can get damaged due to overtwisting if the wind direction keeps on changing and the nacelle keeps on moving in the same direction. Hence, to secure the cable, a counter is present to track the amount of cable twist; it signals the controller against the further movement for cable straightening.

5.4.1.3 The nacelle

A nacelle is the part of the wind turbine that consists of a generator, low- and high-speed shafts, gearbox, brake, and control electronics. It is connected to the tower through a yaw control mechanism. There are two shafts located inside the nacelle, that is, low-speed shaft and high-speed shaft. The rotor turns the low-speed shaft at 30–60 rpm. A step-up gear connects the high-speed shaft to the generator. The gearbox is located in-between these two shafts converting 30–60 to 1200–1800 rpm that is required by the generator to produce electricity. Due to the heavyweight and associated cost, engineers are trying to design gearbox-free directly driven generators capable of generating electricity at low speed.

The main and the heaviest part is the generator that produces 50 or 60 Hz AC power that is carried to the grid utilizing cables. The wind turbine operation can be fixed or variable, depending upon the selection of generator type. Fixed-speed mode is accompanied by synchronous machines to meet the grid frequency requirements. Whereas induction and brushless DC machines are suitable for variable-speed mode operation.

The controlling mechanism of the nacelle contains data related to the wind speed and direction, rotor speed, and generator capacity to enable controlling the parameters of the system. Hence the controller with the help of the yaw mechanism enables the turbine to move along with the wind direction.

5.4.1.4 The tower

The tower is meant to provide the physical strength and support framework to the nacelle and turbine. It is designed in such a way to not only clear the floor but also to provide resistance against vibration caused at higher wind speed. Most of the towers are made of steel, however a concrete base and steel upper are also in use. The height of any tower is associated with the rotor diameter and overall power rating. The height requirements are different for onshore and offshore wind turbines. The water's surface offers no turbulence to wind, hence a lower height of the tower is required. A greater tower height is required for onshore wind turbines to avoid wind turbulence caused by buildings, trees, mountains, hills, etc. Power cables connect turbine generators to the transformer which is located at the lower part of the tower. Some turbines contain a transformer within the nacelle. Whereas power converters in large turbines are located at the base to reduce overall weight. The foundation requirements of onshore and offshore wind turbines are also different and there is a large variety of available options.

5.4.1.5 Control mechanism

The control mechanism in a wind turbine is responsible for keeping the operational range of the wind turbines within certain limits. It can be achieved in an active or passive way depending upon the controlling requirements. They monitor the wind speed, rotor efficiency, electrical and mechanical quantities to perform the required control action. The optimal goal of the control systems is to achieve the optimum power with maximum efficiency under various wind conditions. Power control is applied to some level in wind turbine systems. There are three major ways to control output power. First is stall control that falls under the category of passive controlling technique. The rotor is designed in such a way to self-stall if wind speed exceeds a certain threshold. The second way is regarded as pitch control, which is an active means of controlling the turbines. It causes the turbine blades to turn away from the wind in extreme conditions. It is expensive as a pitching mechanism is required along with a controller to alter the rotor geometry if wind speed goes above or below the threshold values. The third one is the combo of both active and passive forms of controls and is regarded as active stall control. In the case of higher winds, the blades are pitched against the direction to increase the angle of wind attack. Therefore, stalling increases and dynamic wind speed conditions get compensated unlike in the passive controlling mechanism.

5.4.2 Wind turbine classification

There are several factors for classification of wind turbines.

1. Axis position-based classification
2. Size-based classification
3. Speed-based classification
4. Location-based classification

5.4.2.1 Axis position-based classification

Turbines can be classified on the base of axis position as HAWT and vertical axis wind turbines (VAWT). Fig. 5.3 presents the horizontal and vertical axis WTs with essential components.

- *Horizontal axis wind turbines (HAWTs).* HAWTs are commonly used, having a design similar to a windmill. They have a rotor shaft and generator mounted at the top. The rotor shaft is horizontally positioned to capture the maximum wind. These wind turbines have high efficiency and can capture stronger winds. However, they require a massive structure, therefore making installation, maintenance, and transportation difficult. They also requires an additional mechanism that is required to move along the wind and is often subjected to fatigue, stresses, and vibrations.
- *Vertical axis wind turbines (VAWTs).* VAWTs have vertical arrangements of rotor shafts, hence making the turbines self-sufficient against wind directions and additional rotating mechanisms. The generator and other components are required to be installed on the base making maintenance easier. The Darrieus type of wind turbine is the most common in use. These types of turbines have certain advantages, such as they are more economical as they have strong structural support, and yaw control and pitch control is not required. They offer easy installation, maintenance, and transportation. They are suitable and highly efficient in variable and turbulent wind areas. However, they offer lower efficiency, require little push for initial startup, and create noise pollution.

A comprehensive review of HAWTs and VAWTs is presented in terms of the commercial feasibility of novel technologies through the comparative

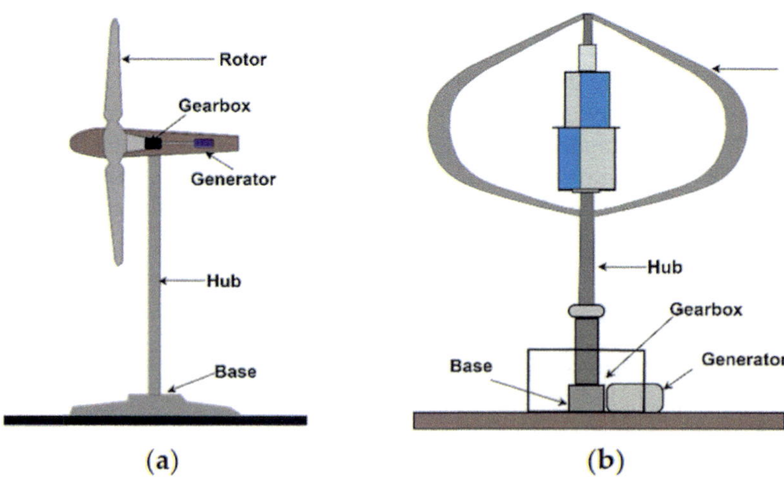

FIGURE 5.3 (A) Horizontal axis wind turbine. (B) Vertical axis wind turbine [35].

analysis of various factors such as power density, efficiency, self-starting ability, and aerodynamics [36,37].

5.4.2.2 Size-based classification

Wind turbines can be classified based on their size as small, medium, and large wind turbines depending upon their power generation capacity. Fig. 5.4 shows the different classification scenarios.

- *Small-scale wind turbines* can be further divided into micro, mini, and household types with the installed power capacity of 0.004–16 kW, blade diameter ranging between 0.5–10 m, and swept area of 0.2–79 m^2. Small commercial wind turbines have an installed capacity ranging between 25–100 kW, blade diameter 10–20 m, swept area of 79–314 m^2, and maximum height up to 50 m. These types are suitable for remote residential areas where the grid is located far away hence making power transmission a difficult and challenging task. However, cost-effectiveness is still in question and research is now focusing on making them more economical and affordable [38].
- *Medium-scale wind turbines* have an installed capacity of 0.1–1 MW, blade diameter 20–50 m, swept area of 314–1963 m^2, and a maximum height of less than 100 m [38]. These types are suitable for remote locations having high demand or commercial buildings. It is common practice to connect these turbines directly to the load through inverters.
- *Large-scale wind turbines* have an installed capacity of 1–3 MW, blade diameter 50–100 m, swept area of 1963–7854 m^2, and a maximum height of 150 m [38]. These types of turbines contain the most complex structure which is meant to contribute toward larger generation capacity.

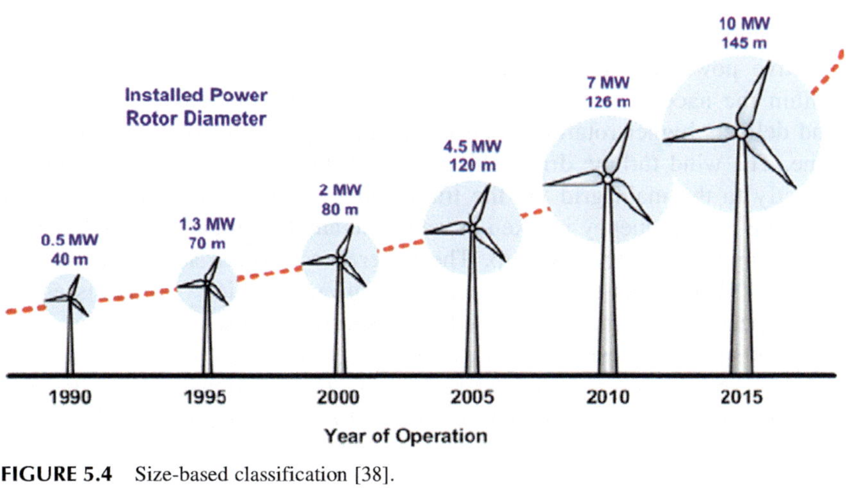

FIGURE 5.4 Size-based classification [38].

Wind farms comprise a few to hundreds of these types connected to the grid. Extreme operating conditions cause these types of turbines to suffer overloading which can lead to damage. Hence, it is inevitable to utilize the latest techniques toward designing a reliable and cost-effective maintenance plan that can successfully monitor the condition and diagnose the faults [39]. There are multiple challenges upon further development of these types of turbines, however, researchers have made significant contributions to point out these and suggested ways to address them [40]. A French-based company, GE Renewable Energy, the fourth largest wind turbine manufacturers in the world in 2018, claimed to have developed the largest wind turbine named "Halide-X" of 13 MW capacity with 220 m rotor diameter and 260 m of total height. They received the first order of 190 machines for Dogger Bank Wind Farm of overall 3.6 GW capacity; it will set a world record and is expected to be completed in 2026.

5.4.2.3 Speed-based classification

Wind turbines can be categorized into fixed-speed wind turbines (FSWT) and variable-speed wind turbines (VSWT).

Fixed-speed wind turbines

FSWTs are the simplest and mosteconomical as compared to their counterparts. During the 1980s and 1990s (the early years of the wind industry), Danish wind turbine manufacturers successfully developed, adopted, and marketed this concept. Therefore it is also regarded as a "Danish concept." FSWTs use a squirrel cage induction generator (SCIG) and multiple stage gearbox to convert mechanical wind power into electrical energy. They are also equipped with the soft starter and capacitor bank to reduce the effect of reactive power consumption. They are mounted on the top of the turbine within the nacelle. The gearbox receives the low rotational speed as input and delivers higher rotational speed within the generator of any specific turbine. The wind turbine drives the generator shaft and its stator is connected directly to the main grid via the transformer. In its normal mode of operation, the grid frequency is taken as reference and the stator frequency is fixed accordingly with very low slip. The difference between stator and slip frequencies determines the rotor speed that offers very low variations against wind speed, hence, they are termed fixed-speed systems.

It is specially designed to achieve the greatest efficiency with a single winding speed, however, in modern systems, two rotational speeds are available to capture a greater amount of wind. It can be obtained by placing two generators in the nacelle or a single generator comprising two freestanding windings, named the generator winding set. It is designed as an eight-pole arrangement to utilize lower wind speeds and four to six pole arrangements for medium and high speeds [41].

FSWTs are not only economical from a maintenance, installation, and transportation perspective but they are also simple, robust, and reliable. However, the high mechanical and fatigue stress, as well as fluctuations in output power, are very common for this type. They offer poor aerodynamic efficiency because the generators cannot be tuned according to the wind speed variations. Power quality control is another issue with these turbines, they require additional equipment for reactive power compensation. FWTs are most used for offshore farms with steadier wind conditions.

Variable-speed wind turbines

VSWTs are specially designed to operate over a wide range of wind speeds. They require a multistage gearbox to boost up the output against very low (variable) wind speeds. The main idea is to capture the maximum amount of available wind which is intermittent by nature. A doubly-fed induction generator (DFIG), and fully rated converter can be used as generators. The variability of the speed leads to the noncompliance of grid frequency requirements, hence special arrangements in the form of back-to-back power converters are required to make sure that the frequency produced is following the standard grid requirements. Generator side converters first convert the upcoming signal to DC (rectification), mainly employed to keep the torque and speed under control through maximum power point tracking. Grid side converters convert the DC to AC signal (inversion); they are used to take care of the reactive power of the grid and upcoming DC voltage. There are various available control techniques and power converter topologies to achieve maximum efficiency. An overview of these techniques can be found in the literature [42].

5.4.2.4 Location-based classification

Wind turbines can be classified into onshore and offshore wind turbines according to their location, as given in Fig. 5.5. Onshore wind turbines can be located at any place having good wind speed characteristics. However, these winds are subjected to various factors as discussed in this chapter previously. Hence, the output power control and protection of equipment are achieved through different controlling schemes.

Offshore wind turbines have the advantage of facing a steady wind speed. A large investment in offshore wind turbines has been observed and this trend is likely to be continued due to various factors including reduced visual impact, reduction in wind turbulence, lower height of towers, and mean speed of the wind is higher. However, there are numerous challenges associated with them. The challenges arise due to the turbine integration with distribution grids. The poor weather conditions restrict their access. Also, per unit electricity cost generation is approximately doubled in the case of offshore turbine systems as compared to the

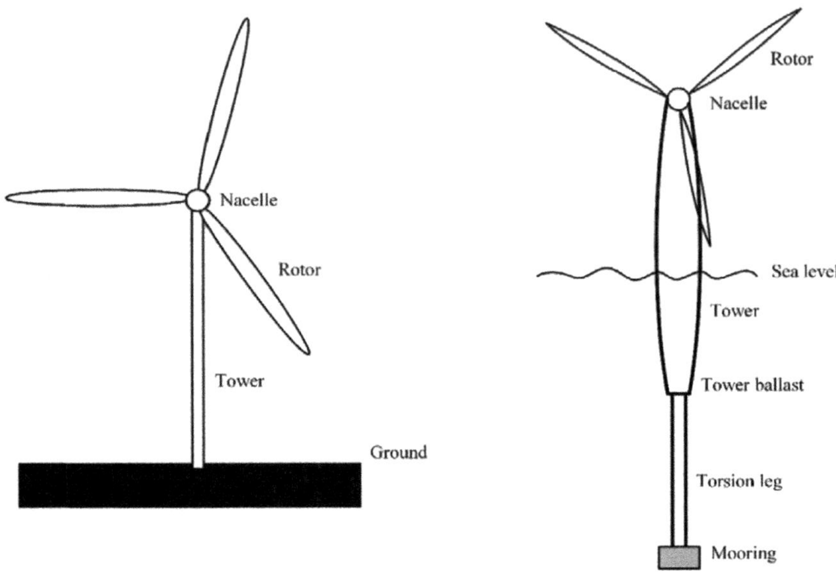

FIGURE 5.5 Onshore WTs (left), offshore WTs (right) [43].

onshore turbines, even steady wind profiles cannot compensate for the high initial costs, expensive submarine cables, and higher operational and maintenance costs.

A newly emerging category of commercial offshore wind turbines has now been under research and development. It is regarded as a floating wind turbine and was installed for the very first time in 2009, off the Norway coast. It is mounted on a floating frame placed in water depths where a fixed structure is impractical. They can increase the available sea area, especially where water depth is low. They have the advantages of offering better fishing accommodations, increasing the available area for shipping lanes, and have access to steady and powerful winds. The first commercial wind turbine of 2.3 MW capacity named Hywind was installed in Norway by Siemens and Statoil. It was installed on a test basis to look for the output and technical issues. But later on, they reported that the floating infrastructure offered no drawbacks. The European Commission reported the largest capacity of floating wind turbines is in Europe, that is, 70% of the world's total. By 2020 it is expected that total installed capacity would reach 210 GW, sharing 14% of the total energy demand and by 2040 it would contribute to 24% of the total energy demand with an expected overall capacity of 350 GW [44].

5.4.3 Generator types

The main purpose of capturing wind is to convert the mechanical power of the wind turbine to electrical energy utilizing a generator. A generator is regarded as a machine and electromechanical component that can be either a DC generator or an AC generator. depending upon the application and type of wind turbine to be used. Electrical machines employ the electromagnetic induction phenomenon that is reversible in a true sense. The opposite machine to the generator is regarded as a motor, that is, capable of converting electrical energy into mechanical energy.

Generators are made up of two major components. The stationary part of the generator is called the "stator" and the rotating part is regarded as the "rotor." Stator and rotor have iron cores and an air gap separates them. A stator has a coil of wires arranged in a special pattern, whereas a rotor may contain an electromagnet or permanent magnet that generates the magnetic field. This magnetic field is a rotating magnetic field that rotates with the rotor. The overall conversion process of electromechanical energy is achieved through mutual interaction of magnetic flux generation in one coil and electrical current in another. The rate of change of flux is directly responsible for the current induced in the coil. The rotating magnetic field produces the voltage in stator terminals. If the stator magnetic field follows the rotating magnetic field of the rotor then it is regarded as a synchronous generator, otherwise called an asynchronous generator.

Wind energy conversion systems (WECS) use a variety of different generators. The main two categories are AC and DC generators, where DC generators are mostly used for low-voltage standalone applications. The most common group of generators employed in wind turbine systems are alternating current generators. These AC generators are further divided into two types, synchronous generators and asynchronous generators. Both have various types used for different types of wind turbines according to the needs and requirements of the project.

5.4.3.1 Synchronous generators

Synchronous generators are most commonly used for power generation. They are widely used in wind turbine systems ranging from kW to MW. They operate at constant or synchronous speed regardless of input torque and following the nominal grid frequency. Hence, they cannot be used in the variable operational mode of wind turbines without the use of proper power electronics converters. They have proven to be a very efficient choice in fixed-speed systems. In grid-connected mode, they enhance the power quality as they don't require reactive power compensation.

The speed of the synchronous generator depends upon the number of rotor pole pairs and rotating field frequency. Synchronous generators generally employ a gearbox to facilitate fixed frequency operations that couple the turbine to the generator. However, multiple machines (a large number of poles) are more efficient and the need for a gearbox gets eliminated. In general, synchronous generators are more complicated and expensive compared to asynchronous generators. The magnetic field in synchronous generators can be produced through the field windings or using permanent magnets. Hence they can be categorized into two main types, wound rotor synchronous generators (WRSG), which use field winding for rotor flux generation, and permanent magnet synchronous generators (PMSG), which use a permanent magnet to produce rotor flux.

- *Wound rotor synchronous generator (WRSG)*: WRSG consists of a wound rotor configuration. It is the backbone of the power industry and hence a well-established machine [45]. Grid-connected stator windings ensure the fixed rotational movement according to the grid frequency. DC excitation source is required to excite the rotor windings using directly connected brushes or brushless exciters. Brushless exciters use rectifiers to convert AC to DC for rotor winding. The advantage of using brushless exciters lies in their easy maintenance but at the cost of high expense and complexity. This DC excitation field in rotor winding runs at synchronous speed.
- *Permanent magnet synchronous generator (PMSG)*: PMSG has a wound stator however the rotor windings are absent and magnetic flux is obtained using a permanent magnet. Hence, they are brushless, lighter in weight, compact in size, highly efficient, and free from issues like thermal stresses and winding losses. But they are very expensive and a complex manufacturing process is associated with them. Also, there exists a threat of demagnetization phenomenon. Excitation through permanent magnets also requires power electronics converters to adjust the frequency and voltage of generation and transmission. A comprehensive review of PMSGs used in grid-connected WECS can be found in the literature [46]. Radial, axial, and transversal flux machines are the most commonly used types. They can be classified into two types depending upon the permanent magnet placement, that is, surface mounted and inset. In surface mounted, PMs are placed on the surface of the rotor separated utilizing non-ferromagnetic material. The inset type uses permanent magnets inside the surface of the rotor, therefore having th higher centrifugal force and greater rotor speed in return.

5.4.3.2 Induction generators

Induction machines are used at a large-scale in the power industry. They are mainly used as motors, however, their simplicity, low cost, low maintenance requirements, excitation without external DC source, and enhanced transient performance make them suitable candidates to be used as generators in small- and large-scale wind turbines. The primary disadvantage associated with these types of generators is the requirement for an external arrangement for a reactive power source for stator magnetization current that is not required in the case of synchronous generators. Hence, it utilizes reactive power through the grid, employing power electronics, capacitor banks, or synchronous generators.

The magnetic field is induced by magnetic saturation in the rotor through the self-excitation phenomenon. The interaction of this field with the stator magnetic field induces an electric field producing rotor winding current due to the relative movement of these fields, regarded as "slip." The extensive use of induction generators in the wind turbine industry lies in the fact that they require little to no maintenance, and possess built-in protection against short-circuit faults, and overloading effects.

There are two types of these generators according to the rotor design, that is, SCIG (short-circuit) and wound rotor induction generator.

- *Squirrel cage induction generator (SCIG)*: it is a very simple, highly efficient machine with no maintenance requirements. Due to the design of the rotor, it is also regarded as a short-circuit rotor machine, as conducting bars of its rotor are placed in the slots, and end rings are employed to short both ends. Hence, no external mechanism is required to control its electrical features. The wind speed changes the slip of the generator, whereas SCIG works very close to its rated speed. Speed deviation is very small against wind speed variations. A major associated drawback is the consumption of reactive power that increases with an increase in wind speed. Hence, to compensate for this effect, it is always associated with the delicate starter mechanism. At the higher speed of the wind, SCIG-based turbines will produce more active power but with the associated requirement of additional reactive power. This is usually provided by the power electronics equipment or capacitor banks, but in their absence this compensation is made by the grid making the overall system unstable. During the grid-connected mode of wind turbines, the transients in the form of inrush current are transmitted directly to the grid due to variations in wind speed. This current may lead to severe voltage disturbance and equipment damage.
- *Wound rotor induction generator (WRIG)*: WRIG is similarly connected to a wind turbine as SCIG with the difference that they include the external mechanism to control electrical characteristics or rotor output. The rotor windings can be connected to power electronics or slip rings and brushes. The use of resistors along with power electronic converters placed inside the rotor eliminates the need for slip rings in the rotor. The variable resistors can quickly control the rotor voltage to maintain continuous power. However, they are more costly than the SCIG and their structure is not as robust and simple. Optically controlled converters can be employed to control the rotor resistance with 10% slip. They are called optic slip IGs.
- Another emerging technology of generator belonging to the same class is regarded as DFIG. As the name suggests it can be supplied/fed from both

ends, that is, rotor and stator. Hence, the rotor and stator can regulate the voltage to control the speed and torque of the generator. The stator windings are directly connected to the main grid and the rotor is connected to the grid through power electronics converters. Hence, it utilizes frequency control instead of resistance control for AC excitation of rotor windings. It can be regarded as a hybrid machine containing characteristics of synchronous and induction generators. It can run like an induction generator but also possess active/reactive power control like synchronous generators. However, in the grid-connected mode of wind turbines, the rotor is decoupled from the grid through converters. Its major advantage lies in a small amount of injected rotor power creating substantial power control in the stator. A variable-speed synchronous generator is required to pass all the power through converters, whereas DFIG needs only 25%–30% of the total power to pass through converters. The remaining 70%–75% passes through stator winding, hence reducing the power electronics rating to 25% of the system. Overall size and associated cost are considerably reduced. But it generates the problem related to the grid-connected mode of operation. There is another disadvantage in that DFIG requires slip rings.

5.4.4 Electrical systems in wind turbines

Wind turbines integrated are regarded as wind farms that are connected to the grid stations to supply electrical energy. A switch breaker isolator is used to disconnect the breaker or whole string. Variable-speed turbines employ AC/AC converters connected by a manual switch. The remote control is used to disconnect the converters from the transformer and reconnect after fault clearance. Fig. 5.6A shows the connection with transformers.

A schematic of the wind farm substation is provided in Fig. 5.6B. The step-up transformers shown are used to convert the generated voltages to higher levels. The equipment on the substation includes bus bars, instrumentation and measurement systems, auxiliary circuits, and protection systems. A current transformer is employed to monitor the individual branch currents.

Lower rated power farms are integrated with some other types of generation units forming an isolated grid. Fig. 5.6C shows the wind–diesel generation where working voltages are of medium type within the range of 10–20 kV. Auxiliary power is delivered to the equipment of the substation and power consumption and quality are monitored [47].

5.4.5 Power electronics integration

Converters are used in wind turbines in various formations and placed between the grid and the WT generator. The selection of these converters depends upon the applications and wind generator types used in the WT

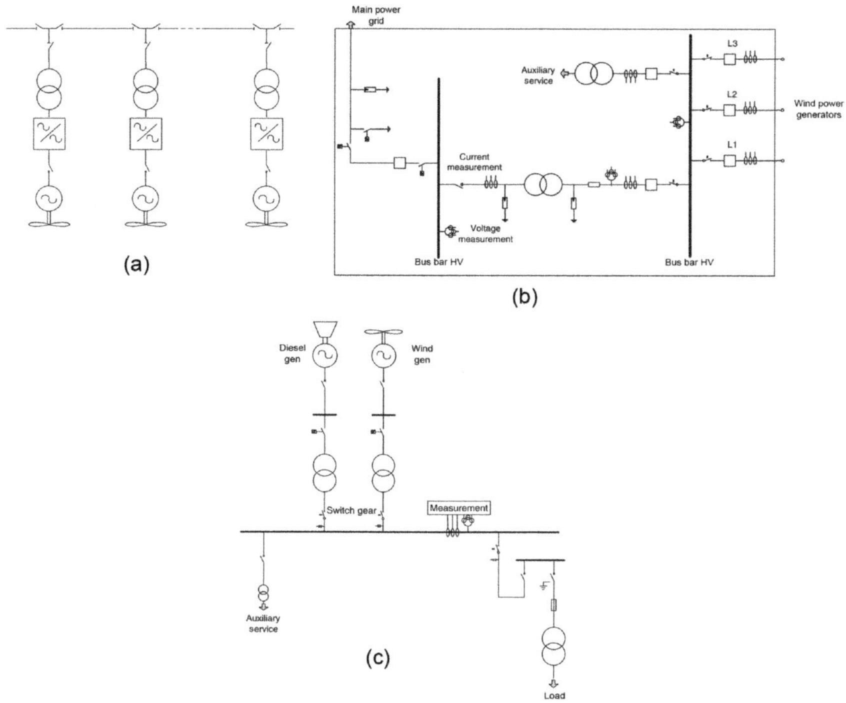

FIGURE 5.6 (A) A common branch shared by multiple turbines [47]. (B) Wind farm substation schematic [47]. (C) Wind–diesel system schematic [47].

system. Generators, after converting the mechanical energy into electrical energy, feed this power to the grid via power electronics converters. These converters are employed in fixed-speed systems to control the inrush current and startup torque. However, in variable-speed systems, these converters are used to control the generator speed and active/reactive grid power. These converters must fulfill the grid and generator requirements. For generators, they must ensure the speed adjustments to extract the maximum power. Whereas for the grid side, converters must satisfy the standard grid requirements, including voltage levels, frequency, and power control (active and reactive). Technology advancements have made the converters more reliable, economical, and capable of handling high voltages and currents efficiently. Therefore empowering the wind turbines to become more energy-efficient, reliable, and cost-effective. Various WECS schemes have been presented in Fig. 5.7.

Some of the power converters used in energy conversion systems of wind turbines are discussed below.

- *Capacitor banks*: the simplistic, easiest, and most economical way to compensate and manage the reactive power of the grid used by induction

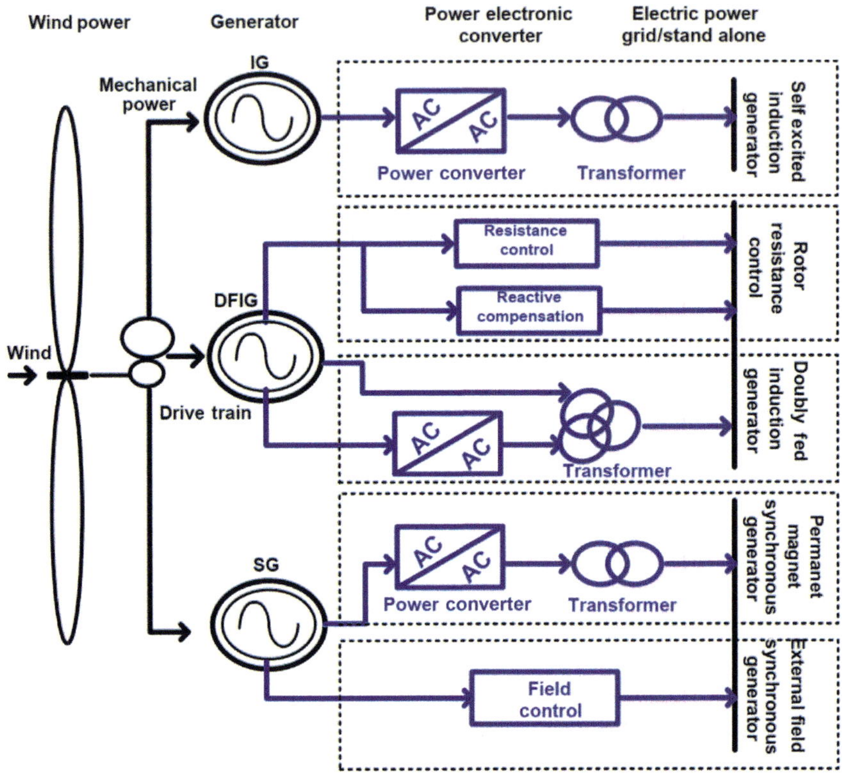

FIGURE 5.7 Various wind energy conversion systems schemes [49].

generators employed in wind turbine systems. The full compensation against dynamic reactive power demand can be achieved by continuous smart monitoring and controlling the number of capacitors over a specific period.

- *Soft starter (AC voltage controller)*: soft starter is used in the wind turbine systems to control the quantity of inrush current. This mechanism enables the generator to start smoothly with less mechanical stress. Therefore the grid disturbances are reduced to a great extent that in return enhances the stability of the grid. They are extensively used with SCIGs and are not suitable for variable-speed operational mode because they are unable to control the output frequency. Once the system becomes operational, soft starter converters are bypassed to eliminate the power losses. In energy conversion systems of wind turbines, silicon-controlled rectifiers (SCRs) are normally used that can be controlled through a firing angle. Firing angle control can adjust the output voltage from zero value to the maximum value (supply voltage), hence, enabling control of the starting current quantity. Without the use of a soft starter converter, the inrush current may grow to seven to eight times the normal rated current [48].

- *Frequency converters*: frequency converters or adjustable speed drivers connect two electrical systems. The main purpose is to control the frequency and voltage in such a way as to enhance the ability of turbines as an active part of power systems. They can convert voltage signal frequency from one level to any other desired level. A general form of frequency converters consists of the rectifier (AC–DC) and inverter (DC–AC) circuits with storage devices (capacitors) in-between them. This process can be accomplished directly or by indirect means. The indirect term refers to the use of DC-link between rectification and inversion processes. Whereas direct conversion is performed without using the DC-link and can be further classified into cycloconverters and matrix converters. Cycloconverters produce the output frequency less than the input frequency and matrix converters can provide output frequency less, equal, or greater than the input frequency [49].

5.4.6 Economics

Wind energy is considered to be among the cheapest electrical energy sources. Over the past few decades, the Wind turbine industry has observed a steady decline in overall cost. Several factors integrate to form the overall cost including site, installation, maintenance, machinery, and grid transmission. It requires a higher initial investment with 80% of the capital cost relating to the machinery. The advancement in technology is enabling the wind industry to produce machines of greater ratings and associated cost is dropping down. It is a very efficient choice in terms of zero fuel requirements with nominal operational and maintenance costs. To produce the same power, a coal plant requires five times more land amount as compared to a wind farm. Hence, to extract green, cheap, continuous, and stable electricity, investing in a wind energy project is an efficient choice. However, an extensive economic review is necessary and all associated factors must be considered to reach a good decision regarding establishing a wind power plant project [50].

The per-unit cost can be easily calculated with the help of capital investment and operational cost. the four major cost factors associated with the wind power project are capital cost, variable cost, operational cost, and life-cycle cost. where each of them can be further divided into various subcategories. Wind power electricity cost is defined by the following factors [51]:

1. Equipment depreciation
2. Borrowed capital interest rate
3. Maintenance and operational cost
4. Local and federal taxes

5. Government policies, incentives
6. Royalty payments
7. Electricity consumption on standby mode
8. Energy storage devices

A comprehensive economic analysis of wind energy cost can be found in the literature [51].

Another associated factor is the installation place. As there is a huge difference in the cost of onshore and offshore wind turbines. The expense related to offshore wind turbines is greater than the onshore turbines due to the factor of the higher initial setup cost and expensive operational and management costs. However, Europe in 2019 added 10 new wind farms including 502 offshore grid-connected turbines of 3.623 GW capacity. Taking overall offshore capacity to 22.072 GW with 5047 grid-connected turbines in 12 countries [52]. The researchers have suggested ways to decrease the generation cost of wind energy. The improvement in research and development practices, optimization of turbine size, advancements in blades materials, enhancing forecasting techniques, site selection procedure, optimization in operational and management techniques, removing administrative hurdles, and easy payment procedure must be adopted [53].

5.5 Reliability science of wind turbines

Reliability is the probability of a device performing its required function with satisfactory output under the predefined condition for a specific time frame. Reliability is the measure of determining the failure rate of any product using historic data, working environmental conditions, analytical techniques, and experimental evidence. Researchers work on these issues to develop robust devices that can tackle these extreme conditions. Modern wind turbines are now made up of complex infrastructure containing various hardware- and software-supported components. It is now inevitable to consider the reliability factors in wind turbine setups. The operational and maintenance cost can be reduced using larger output generating wind turbine units but still cost per failure is observed to be increasing. This cost can be further reduced by enhancing the reliability of wind turbines.

The wind turbine industry has now implemented condition monitoring systems (CMS) for enhanced maintenance and repair strategies. CMS can be used for the individual level (wind turbine units) as well as cohort level (wind farms). They are used to monitor the monitoring of the critical components including generator, blades, gearbox, bearings, and tower. These systems in conjunction with fault detection schemes are proven to be extremely effective and provide an early warning related to electrical and mechanical faults. Hence, making it possible to schedule the yearly maintenance plan, which is extremely important for offshore wind turbines. It also reduces the inspection visits, hence reducing overall maintenance cost. A productive maintenance strategy can benefit in several ways [54]:

- Avoidance of catastrophic failures, defects, and premature breakdowns
- Maintenance cost reduction
- Remote access to supervise and detect faults for prescheduled maintenance activity
- Data collection for wind turbine behavior for future developments
- An informed operator may enhance the decision-making process and operational efficiency

There are different perspectives on wind turbine taxonomy. It helps in data collection of wind turbines for reliability purposes. ReliaWind consortium is a five-level system to describe wind turbines [55]:

- *A system* (wind plant), comprising turbines, substations, and different types of cables
- *A subsystem*, an independent turbine, or any substation at the plant
- *An assembly*, comprising wind turbine gearbox or HV substation
- *Subassembly*, moving shaft or rotor in the gearbox
- *Component*, bearing of the shaft

These parts contribute to delivering the required output. Each of the parts in a wind turbine must perform its function, the wind turbine will fail to deliver otherwise. There are mainly four types of failure events that are faced with wind turbines. *Infant failure* events associated with the product quality; *premature failures* are related to design defects, manufacturing defects, or operational and maintenance defects; *random failures* that can be expected throughout the lifecycle due to unknown events; and *wear-out failures* are the ideal ones in which failures occur after completion of the expected lifecycle. A list of various failure types, their failure mechanism, modeling, and analysis are given by researchers [56]. In the current scenario, a wind farm may include hundreds of wind turbines, hence provide a huge amount of data. This data needs to be collected, analyzed, evaluated, and interpreted to get an insight into how to enhance the reliability and availability of wind turbines. Some other factors govern the lifecycle management of various components. These factors include modeling fault diagnosis and failure predictions.

There are some challenges in pursuing reliable operations of wind turbine systems. Many researchers have adopted a rigorous approach in highlighting these challenges and presented a comprehensive review of operational and maintenance initiatives for wind turbines [57,58].

- Absence of a standard data collection format
- Unavailability of critical data
- No standard procedure to follow for commissioning and operation of wind turbines

- Adaptation of different technologies by different manufacturers hence making the relative comparison a difficult task
- It is difficult to compare new and old wind turbines due to the advancements in technologies
- Relative comparison is also a hard task in terms of different weather environments under which turbines are installed
- These systems are new to the world, and technological advancements are quickly occurring. Hence, there is a lack of available literature on emerging technologies, as their lifecycle of 20 years is not yet completed. Whereas available fault diagnosis and analysis for old turbines do not apply to newer technology-based turbines

Critical assembly identification is proven to be the most effective in contributing to CMS. There is a need for a WT taxonomy-based standard database due to a lack of literature on the reliability and availability of wind turbine systems. Normalization of the data under a common taxonomy has been performed and analysis and comparison of different reliability studies can be concluded as [59]:

1. The data required for CMS must be taken from the most critical parts of wind turbines, including gearbox, generators, hub, turbine blades, control, and electric systems.
2. Structures and sensors are found to be the most reliable parts with the lowest failure rates.
3. Dispersion rates for failure and downtime are higher for critical parts of wind turbines compared to other parts.
4. Control systems and electric systems are the top contributors to failure rates.
5. Gearbox, hub, and blades are the top contributors in downtime.
6. If mechanical and electrical components are compared, they show the same failure rates, however downtime of mechanical assemblies is greater than electrical ones, that is, 75% of the overall downtime.
7. Recent studies have shown a significant improvement in the reliability of various WT assemblies, such as shaft, hydraulic, braking, yaw, and sensor. Whereas the gearbox and generator have remained the same.
8. Thirty-eight percent of the total assemblies are responsible for 50% of the WT failure rates and 60% of overall downtime.
9. The gearbox is the most critical assembly responsible for the most failures.

The deficiency of reliability data was found to be insufficient. However, researchers have suggested developing a public database that must include operational and maintenance costs. it will give a deep insight into the failure reasons, for better design and CMS development. Hence, the real scenario-based maintenance mechanism can be applied and scheduled instead of applying conventional techniques. It will not only improve the failure rates to enhance the availability of WTs, but also will bring down the associated cost [59].

5.6 Energy storage options of wind turbines

Energy storage is regarded as the most vital tool to preserve renewable energy. Technology advancements have made it now possible to use advanced storage options, thus limiting the requirements of grid infrastructure. Wind energy penetration into the power systems requires a sophisticated backup storage mechanism due to the dynamic and nondispatchable nature of wind energy. So many options for energy storage are available now. The parameters associated with the storage units are *power storage capacity* related to maximum available power, *energy storage capacity* referring to maximum energy storage ability, *response time* describing the time required to start delivering power, efficiency, and charge to discharge ratio.

The overall energy storage unit comprises three main parts. First storage means are required to retain potential energy in the storage device in the form of electrical, mechanical, or chemical energy. Second, a power electronics interface is required for the conversion process if required. Rectifiers are required to convert the AC to DC for charging the storage device and then inverters are employed to convert DC back to AC if power is required to be delivered from the storage device. Whereas charging, discharging, and standby constraints are facilitated to avoid any form of damage to the device. Third, the balance of the plant includes miscellaneous components including protection devices, electricity connections, supporting rack, environmental controlling mechanism, roadways, and all paperwork activities. Energy storage devices can be categorized under different power scales and storage capacities [60].

- Large-scale storage capacity
  - Pumped hydroelectric energy storage (PHES)
  - Underground PHES
  - Compressed air energy storage
- Medium-scale storage capacity
  - Battery energy storage
  - Flow-battery energy storage
  - Supercapacitor energy storage
  - Flywheel energy storage
  - Superconducting magnetic energy storage
- Small-scale storage capacity
  - Hydrogen energy storage systems
  - Thermal energy storage systems
  - Electric vehicles

The selection of a particular device depends upon the application and requirements of the system because there exists no device with all the characteristics in a single technology.

5.7 Application of wind turbines

Applications of wind turbines can be given as:

- *On-grid applications*: on-grid applications of wind turbines usually are of MW capacity. A large number of turbine units are bought together to develop a wind farm. This trend is increasing nowadays. The land of these farms can be used also for some other useful means, such as agriculture purposes.
- *Off-grid applications*: remote locations are the best suitable places for wind turbine applications. As transportation of electricity becomes too costly especially for less power consumption requirement, that is, <10 kW. The basic purpose of these wind turbines is to store the energy employing some storage devices and supply the energy when and if required.
- *Wind pumps*: wind turbines are also suitable for supplying energy to water pumps. These water pumps are meant to store the water rather than any form of energy. Hence additional requirements of energy storage devices are eliminated and water is saved for future use [61].
- *Hybrid systems*: electricity production in remote areas is expensive as fuel transportation is continuously required by some means. However, depending on the wind conditions and feasibility studies, wind turbines can share the overall electricity demand. Hybrid systems involving wind turbines are normally wind–diesel and PV–wind hybrid energy systems. The key role involves the reduction in fossil fuel (diesel) consumption in the case of wind–diesel hybrid systems [61]. Whereas PV–wind hybrid systems are well-known due to their extensive experimental and practical use. A handy monthly inverse relationship is observed among these two quantities, during the times of higher solar irradiance, wind speeds are found to be lower. On the other hand, months of higher wind speeds observe a lower solar irradiance amount.

Problems

Problems 1–5 contain three/four answer options A, B, C, and D. Choose the correct answer.

1. What is the main source for the formation of wind?
   A. Uneven land
   B. Sun
   C. Vegetation
   D. Seasons

2. What happens when the land near the Earth's equator is heated?
   A. All the oceans gets heated up
   B. Small wind currents are formed
   C. Rise in tides
   D. Large atmospheric winds are created
3. What are used to turn wind energy into electrical energy?
   A. Turbine
   B. Generators
   C. Yaw motor
   D. Blades
4. Wind energy is harnessed as ______ energy with the help of a windmill or turbine.
   A. Mechanical
   B. Solar
   C. Electrical
   D. Potential
5. Turbines are classified as
   A. Axis position-based classification
   B. Size-based classification
   C. Speed-based classification
   D. All of the above
6. Define different types of speed associated with wind energy.
7. On which factors do wind power depends. Briefly explain your answer.
8. What are the main components of a wind turbine?
9. Which factors are involved in the economics of wind power?
10. Write down the four applications of wind turbines.

References

[1] T.J. Price, James Blyth—Britain's first modern wind power pioneer, Wind. Eng. 29 (3) (2005) 191–200.

[2] M.J. Pasqualetti, R. Righter, P. Gipe, Wind energy, history of, in: C.J. Cleveland (Ed.), Encyclopedia of Energy, Elsevier, New York, 2004, pp. 419–433.

[3] Danish Wind Power Association.<http://www.windpower.org/en>.

[4] N.E. Selin, Wind power. <https://www.britannica.com/science/wind-power>, 2020.

[5] WWEA, World wind capacity at 650,8 GW, Corona crisis will slow down markets in 2020, renewables to be core of economic stimulus programmes. <https://wwindea.org/blog/2020/04/16/world-wind-capacity-at-650-gw/>, 2020.

[6] IRENA, Future of wind: deployment, investment, technology, grid integration and socio-economic aspects (A Global Energy Transformation paper), Abu Dhabi, 2019. <https://www.irena.org/-/media/Files/IRENA/Agency/Publication/2019/Oct/IRENA_Future_of_wind_2019.pdf>, 2019.

[7] V. Fthenakis, H.C. Kim, Life-cycle uses of water in U.S. electricity generation, Renew. Sustain. Energy Rev. 14 (7) (2010) 2039–2048.

[8] M.S. Nazir, A.J. Mahdi, M. Bilal, H.M. Sohail, N. Ali, H.M.N. Iqbal, Environmental impact and pollution-related challenges of renewable wind energy paradigm—a review, Sci. Total. Environ. 683 (2019) 436–444.

[9] D.K. Clarke, Wind turbine noise: the facts and the evidence. <https://ramblingsdc.net/wtnoise.html>, 2012.

[10] M.S. Nazir, N. Ali, M. Bilal, H.M.N. Iqbal, Potential environmental impacts of wind energy development: a global perspective, Curr. Opin. Environ. Sci. Health 13 (2020) 85–90.

[11] P. Hartmann, V. Apaolaza-Ibáñez, Consumer attitude and purchase intention toward green energy brands: the roles of psychological benefits and environmental concern, J. Bus. Res. 65 (9) (2012) 1254–1263.

[12] M.J. Coren, Germany had so much renewable energy on Sunday that it had to pay people to use electricity. <https://qz.com/680661/germany-had-so-much-renewable-energy-on-sunday-that-it-had-to-pay-people-to-use-electricity/>, 2016.

[13] IEA_Wind. <https://community.ieawind.org/home>, 2016.

[14] <http://janmun.com/wip/research.html>.

[15] A. Pearson, Safety fears as turbines toppled. <http://www.thejournal.co.uk/news/north-east-news/safety-fears-as-turbines-toppled-4511322>, 2008.

[16] R. Oblack, Synoptic scale vs. mesoscale weather systems. <https://www.thoughtco.com/synoptic-scale-vs-mesoscale-weather-systems-3444176>, 2020.

[17] A.M. Helmenstine, What is the density of air at STP? <https://www.thoughtco.com/density-of-air-at-stp-607546>, 2020.

[18] Where does wind energy come from? <http://xn--drmstrre-64ad.dk/wp-content/wind/miller/windpower%20web/en/tour/wres/index.htm>.

[19] G.M. Masters, Renewable and Efficient Electric Power Systems, Wiley, New York, 2004.

[20] New York State Energy Research and Development Authority, Wind Resource Assessment Handbook. <http://www.nyserda.ny.gov/-/media/Files/Publications/Research/Biomass-Solar-Wind/wind-resource-assessment-toolkit.pdf/>, 2020.

[21] M. De Lellis, R. Reginatto, R. Saraiva, A. Trofino, The betz limit applied to airborne wind energy, Renew. Energy 127 (2018) 32–40.

[22] R.D. Prasad, R.C. Bansal, Technologies and Methods Used in Wind Resource Assessment, Handbook of Renewable Energy Technology, World Scientific, Singapore; Hackensack, NJ, 2011.

[23] Y. Jang, E. Byon, Probabilistic characterization of wind diurnal variability for wind resource assessment, IEEE Trans. Sustain. Energy 11 (4) (2020) 2535–2544.

[24] Y.H. Juan, C.Y. Wen, W.Y. Chen, A.S. Yang, Numerical assessments of wind power potential and installation arrangements in realistic highly urbanized areas, Renew. Sustain. Energy Rev. 135 (2021) 110165.

[25] B. Josimović, A. Cvjetić, D. Furundžić, Strategic environmental assessment and the precautionary principle in the spatial planning of wind farms—European experience in Serbia, Renew. Sustain. Energy Rev. 136 (2021) 110459.

[26] R. Rabbani, M. Zeeshan, Exploring the suitability of MERRA-2 reanalysis data for wind energy estimation, analysis of wind characteristics and energy potential assessment for selected sites in Pakistan, Renew. Energy 154 (2020) 1240–1251.

[27] J. Waewsak, C. Kongruang, Y. Gagnon, Assessment of wind power plants with limited wind resources in developing countries: application to Ko Yai in southern Thailand, Sustain. Energy Technol. Assess. 19 (2017) 79–93.

[28] M. Irfan, Y. Hao, M.K. Panjwani, D. Khan, A.A. Chandio, H. Li, Competitive assessment of South Asia's wind power industry: SWOT analysis and value chain combined model, Energy Strategy Rev. 32 (2020) 100540.

[29] F. Noman, et al., Multistep short-term wind speed prediction using nonlinear auto-regressive neural network with exogenous variable selection, Alex. Eng. J. (2020).

[30] W. Li, X. Jia, X. Li, Y. Wang, J. Lee, A Markov model for short term wind speed prediction by integrating the wind acceleration information, Renew. Energy 164 (2021) 242–253.

[31] W. Xu, et al., Multi-step wind speed prediction by combining a WRF simulation and an error correction strategy, Renew. Energy 163 (2021) 772–782.

[32] Y. Zhang, S. Gao, J. Han, M. Ban, Wind speed prediction research considering wind speed ramp and residual distribution, IEEE Access. 7 (2019) 131873–131887.

[33] Q. Zhu, et al., Learning temporal and spatial correlations jointly: a unified framework for wind speed prediction, IEEE Trans. Sustain. Energy 11 (1) (2020) 509–523.

[34] T. Ulery, M. Riley, Increasing wind turbine reliability through blade pitch control upgrades, USA. <https://www.power-eng.com/2020/01/17/increasing-wind-turbine-reliability-through-blade-pitch-control-upgrades/#gref>, 2020.

[35] K. Shivam, J.-C. Tzou, S.-C. Wu, Multi-objective sizing optimization of a grid-connected solar–wind hybrid system using climate classification: a case study of four locations in Southern Taiwan, Energies, MDPI 13 (10) (2020).

[36] J. Liu, H. Lin, J. Zhang, Review on the technical perspectives and commercial viability of vertical axis wind turbines, Ocean. Eng. 182 (2019) 608–626.

[37] A. Das, K.B. Chimonyo, T.R. Kumar, S. Gourishankar, C. Rani, Vertical axis and horizontal axis wind turbine: a comprehensive review, in: 2017 International Conference on Energy, Communication, Data Analytics and Soft Computing (ICECDS), 2017, pp. 2660–2669.

[38] A. Tummala, R.K. Velamati, D.K. Sinha, V. Indraja, V.H. Krishna, A review on small scale wind turbines, Renew. Sustain. Energy Rev. 56 (2016) 1351–1371.

[39] Z. Liu, L. Zhang, A review of failure modes, condition monitoring and fault diagnosis methods for large-scale wind turbine bearings, Measurement 149 (2020) 107002.

[40] R. McKenna, P. Ostman, vd Leye, W. Fichtner, Key challenges and prospects for large wind turbines, Renew. Sustain. Energy Rev. 53 (2016) 1212–1221.

[41] M. Orabi, F. El-Sousy, H. Godah, M.Z. Youssef, High-performance induction generator-wind turbine connected to utility grid, in: INTELEC 2004, 26th Annual International Telecommunications Energy Conference, 2004, pp. 697–704.

[42] O. Apata, D.T.O. Oyedokun, An overview of control techniques for wind turbine systems, Sci. Afr. 10 (2020) e00566.

[43] S. Wang, S. Wang, J. Liu, Life-cycle green-house gas emissions of onshore and offshore wind turbines, J. Clean. Prod. 210 (2019) 804–810.

[44] Why the EU supports wind energy research and innovation. <https://ec.europa.eu/info/research-and-innovation/research-area/energy-research-and-innovation/wind-energy_en#:~:text=The%20EU%20currently%20has%20the,to%2024%25%20of%20electricity%20demand>, 2020.

[45] S. Nuzzo, M. Galea, C. Gerada, N. Brown, Analysis, modeling, and design considerations for the excitation systems of synchronous generators, IEEE Trans. Ind. Electron. 65 (4) (2018) 2996–3007.

[46] S.M. Tripathi, A.N. Tiwari, D. Singh, Grid-integrated permanent magnet synchronous generator based wind energy conversion systems: a technology review, Renew. Sustain. Energy Rev. 51 (2015) 1288–1305.

[47] J.M. Carrasco, E. Galván, R. Portillo, Chapter 4—Wind turbine applications, in: M.H. Rashid (Ed.), Alternative Energy in Power Electronics, Butterworth-Heinemann, Boston, 2011, pp. 177–230.

[48] R.W. Erickson, D. Maksimovic, Fundamentals of Power Electronics, second ed., Springer US, 2001, p. 883.

[49] K.B. Tawfiq, A.S. Mansour, H.S. Ramadan, M. Becherif, E.E. El-kholy, Wind energy conversion system topologies and converters: comparative review, Energy Proc 162 (2019) 38–47.

[50] T. Wizelius, J. Earnest, Wind Power Plants and Project Development, second ed., PHI Learning Private Ltd, New Delhi, 2017, p. 532.

[51] M. Ragheb, Chapter 25—Economics of wind power generation, in: T.M. Letcher (Ed.), Wind Energy Engineering, Academic Press, 2017, pp. 537–555.

[52] L. Ramírez, D. Fraile, G. Brindley, Offshore wind in europe, key trends and statistics 2019, Brussels, Belgium. <https://windeurope.org/about-wind/statistics/offshore/european-offshore-wind-industry-key-trends-statistics-2019/>, 2020.

[53] M.I. Blanco, The economics of wind energy, Renew. Sustain. Energy Rev. 13 (6) (2009) 1372–1382.

[54] F.P.G. Márquez, J.M. Pinar Pérez, Chapter 1—Wind turbines: a general reliability analysis, in: M. Papaelias, F.P.G. Márquez, A. Karyotakis (Eds.), Non-Destructive Testing and Condition Monitoring Techniques for Renewable Energy Industrial Assets, Butterworth-Heinemann, Boston, 2020, pp. 1–18.

[55] P. Tavner, Offshore Wind turbines: reliability, availability and maintenance (Energy Engineering), The Institution of Engineering and Technology (2012) 296.

[56] E.E. Kostandyan, Reliability modeling of wind turbines: exemplified by power converter systems as basis for O&M planning, Ph.D., Department of Civil Engineering, Aalborg University, Aalborg, Denmark, PhD Thesis, 2013.

[57] E. Echavarria, B. Hahn, G.J.W. van Bussel, T. Tomiyama, Reliability of wind turbine technology through time, J. Sol. Energy Eng. 130 (3) (2008).

[58] S. Pfaffel, S. Faulstich, K. Rohrig, Performance and reliability of wind turbines: a review, Energies 10 (1904) (2017) 27.

[59] E. Artigao, S. Martín-Martínez, A. Honrubia-Escribano, E. Gómez-Lázaro, Wind turbine reliability: a comprehensive review towards effective condition monitoring development, Appl. Energy 228 (2018) 1569–1583.

[60] W. Tong, Wind Power Generation and Wind Turbine Design, WIT Press, Boston, MA, 2010, p. 769.

[61] Introduction to clean energy project analysis, in: Clean Energy Project Analysis: Retscreen® Engineering & Cases Textbook, Canada. <http://www.alternativesynergy.org/SEDV-work/Project/RETScreen/Textbook_Intro.pdf>, 2005.

Chapter 6

Hydro energy

Muhammad Kamran
Department of Electrical Engineering and Technology, Riphah International University, Pakistan

6.1 Introduction

Hydropower is another indirect form of solar energy. It is considered the most mature and environment-friendly renewable energy resource. Hydropower is the renewable energy obtained by water falling from high potential to low potential. The hydro energy is harnessed by using the potential energy of the falling water by running a hydro turbine. This turbine is coupled to the rotor of the electrical generator. A set of three-phase voltages is induced in the stator of the electrical generator.

The water cycle depends on the sun's energy. When sunlight falls on the surface of water in rivers, lakes, and seas, the upper surface molecules get warmer. When they get enough energy, they leave the surface and escape into the atmosphere, via the evaporation process. These molecules get cooler when they rise up into the atmosphere. When these molecules pass through the cold areas where the air is cool, they condensate and become water droplets. As the droplet holding capacity of the cold air is lower than the warm air, these water droplets cause the rain and maintain the circulation of the water in the atmosphere. If it rains in the mountain ranges, it becomes ice and makes glaciers. If it rains in the fields, somehow or other it finds its way to the sea and the cycle starts again.

6.2 Basic components of the hydropower plant

6.2.1 Dam

A dam is a barrier structure that is built across a river to retain the water for the large storage of water. This water storage releases the water primarily for hydropower and secondarily for irrigation purposes [1].

Renewable Energy Conversion Systems. DOI: https://doi.org/10.1016/B978-0-12-823538-6.00007-5

6.2.2 Penstock

A penstock is a pipe that channels the water from the dam to the turbine. On the dam side valves or gates are used to control the flow of the water to the turbine. The penstock could be a cemented, polyvinyl chloride (PVC), or a steel pipe with a specific length, diameter, and thickness properly calculated for the available flow rate and the head.

6.2.3 Turbines

A turbine is the key part of any hydropower system as it converts the potential energy of the water into the mechanical rotational energy that is used to rotate the rotor of the electrical generator. The type of turbine that would be suitable for a hydropower plant depends upon the flow rate and the capacity of the power plant. Only Pelton turbines are suitable for the large heads. Francis turbines can be used in the middle area from around 500 m. In the area where the head overlaps with the Pelton turbines of around 100–500 m, the Pelton turbine is more insensitive to wear, but the Francis turbine is superior due to its higher efficiency, better application possibilities with fluctuating heads, and smaller dimensions. Kaplan turbines are used for the smaller heads. Here, too, there is an overlap with the Francis turbines in the range of around 40–50 m. So, the Kaplan turbines are usually preferred here. The types of turbines used in a hydropower plant are shown in Fig. 6.1.

6.2.3.1 Impulse turbines

Impulse turbines consisting of buckets on the runner use the kinetic energy of the water stream. The flowing water after hitting the buckets and rotating the runner leaves the housing of the turbine. The impulse turbines find their best applications in high head and low flow rate sites. Pelton, Turgo, and cross-flow turbines are types of impulse turbines [2].

Pelton turbines are best for the high head water drops and low flow rate. These turbines consist of a circular disk mounted on the shaft coupled to the

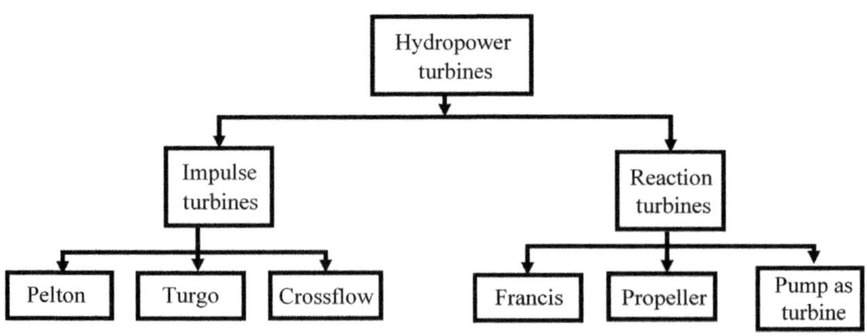

FIGURE 6.1 Types of hydropower turbines.

FIGURE 6.2 Pelton turbine.

rotor of the generator. The buckets are mounted along the circumference of the circular disk in such a way that when water hits the buckets they move along the tangent of the disk, as shown in Fig. 6.2. The efficiency of the Pelton turbine is restricted to 90% by the friction losses, nozzle losses, and aerodynamic drag. As the load on the turbine changes, the speed of the turbine varies accordingly, which changes the frequency of the electricity. To maintain the frequency and the speed of the turbine, the flow rate through the turbine is controlled by the valve mounted on the jet.

6.2.3.2 *Reaction turbines*

Reaction turbines are the turbines that use the pressure as well as the velocity of the moving water to rotate. Reaction turbines are placed in the water stream where the water enters the casing tangentially. After rotating the blades the water axially leaves the casing of the turbine. Francis and propeller turbines are the types of reaction turbines [3].

The Francis turbine, as it would appear from the name, was developed by the US engineer James Francis. For its proper operation, the turbine is submerged in water in such a way that water enters along the radial axis and after rotating the turbine blades it leaves the turbine vertically through the center of the turbine. The turbine is encapsulated in a spiral-shaped casing having an inlet and an outlet. The shaft of the turbine is coupled to the electrical generator that produces electricity, as shown in Fig. 6.3. The disadvantage of the

FIGURE 6.3 Francis turbine coupled with an electrical generator.

Francis turbine is its inability to operate efficiently under varying conditions. Propeller and Kaplan turbines are similar in the head, flow rate, and physical appearance. Their working principle is similar to the Francis turbine. Among all turbines, they are best for the high flow rate and the low head potential.

6.2.3.3 *Tailrace*

The water coming from the penstock transfers its pressure to the turbine and finds its way out of the turbine downstream. Downstream the water returns to its original path.

6.2.3.4 *Electric generators*

Electrical generators are coupled with the hydro turbines that convert the rotational energy of the turbine into electrical energy. They work on the principle of the faraday law of electromagnetic induction, which states that when the flux through a closed loop of the conductor is changed concerning time, a voltage is developed across the conductor. An electrical generator consists of a rotor and a stator. A field current is provided to the rotor conductor generating magnetic flux. When the rotor is rotated by the prime mover (hydro turbine in hydropower), its flux is changed in the stator conductors that produce a set of three-phase voltages in the stator. Eq. (6.1) is the Faraday law of electromagnetic induction.

$$E = -N.\frac{d\varnothing}{dt} \tag{6.1}$$

The electric frequency of the developed voltages in the stator depends upon the synchronous speed and the number of poles, as shown in Eq. (6.2). As the number of poles is constant, the frequency of the generated voltage is dependent on the speed. The flow and the speed of the water falling on the turbine continuously change, bringing fluctuations in the electrical frequency. To overcome this problem, variable frequency drives (VFDs) are used. In VFDs, variable alternating current (AC) signal is converted into direct current and then again into constant frequency AC power.

$$N_s = \frac{120 f_e}{P} \tag{6.2}$$

Synchronous generators are widely used electrical generators in hydropower plants. These generators are synchronized with the grid by matching the line voltages, phase sequence, phase angle, and frequency.

Example 6.1

The conductor coil of the synchronous generator on the pole is 0.5 m on each side producing a magnetic field of 2 T. On increasing the load, flux increases from 2 T to 3 T in 15 seconds. the coil contains 10 turns. Calculate the induced voltage.

Solution

The induced voltage is determined by Faraday's law of electromagnetic induction, given in Eq. (6.1).

$$E = -N \cdot \frac{d\varnothing}{dt}$$

$$E = -N \cdot \frac{\varnothing_f - \varnothing_i}{dt}$$

$$E = -N \cdot \frac{B_f A - B_i A}{dt}$$

$$E = -N \cdot \frac{A(B_f - B_i)}{dt}$$

$$E = -10 \cdot \frac{0.25(3-2)}{15}$$

$$E = 0.167\ V$$

Example 6.2

An electric generator coupled with the micro hydro turbine consists of two poles generating an electrical frequency of 60 Hz. What would be the synchronous speed of the generator?

Solution

Eq. (6.2) is used to determine the synchronous speed of the synchronous generator coupled to the hydro turbine.

$$N_s = \frac{120f_e}{P}$$

$$N_s = \frac{120(60)}{2}$$

$$N_s = 3600 \text{ rpm}$$

The electric frequency of any electric power system remains constant. So by changing the number of poles of the generator, we can change the synchronous speed/rotation of the synchronous generator. Table 6.1 shows the synchronous speed of the generator for various numbers of poles at 50 Hz and 60 Hz frequency of the electric system.

6.3 Small/micro hydropower

Following the pattern of the large hydropower plants, the design and layout of the small and micro hydropower plants depend upon the size of the project. Generally, a hydropower plant larger than 100 MW is considered a large hydropower plant. A power plant less than 100 kW is considered a micro hydropower plant. Some researchers divide the micro hydro into pico and micro hydro. A plant between 5 and 100 kW is micro hydro whereas a plant less than 5 kW is called a pico hydropower plant. Similarly, a power plant larger than 100 kW and less than 2 MW is considered as a mini hydropower plant, whereas if the range is between 2 and 25 MW it is categorized as a small hydropower plant. A hydropower project between 25 and 100 MW is categorized as medium hydro. The details of the capacity range and the corresponding category are given in Table 6.2.

The design of the civil construction, as well as the turbine and generator of each category given in the above table, is designed specifically for that

TABLE 6.1 Synchronous speed of generators at different poles and frequency.

Sr. no.	No. of poles	N_s at 50 Hz (rpm)	N_s at 60 Hz (rpm)
1	2	3000	3600
2	4	1500	1800
3	6	1000	1200
4	8	750	900

TABLE 6.2 Types of hydropower plants concerning the capacity range.

Sr. no.	Type of hydro	Capacity range
1	Large hydro	>100 MW
2	Medium hydro	25,001 kW–100 MW
3	Small hydro	2001 kW–25 MW
4	Mini hydro	101 kW–2 MW
5	Micro hydro	5–100 kW
6	Pico hydro	<5 kW

project depending upon the availability of water, load demand, and the type of terrain. The design that is common among all categories is given in Section 6.4.

6.4 Designing of the small/micro hydropower system

6.4.1 Flow duration curve

One method of measuring the flow rate of water in the mainstream is the float method. In this method take a realistic distance in the open channel and mark upstream and downstream positions on the same bank of the stream. Note down the time taken by a floating object (preferably wood) to travel from the specified upstream to the specified downstream. It is better to repeat the process at least three times and average them [1]. The velocity of the water is determined by dividing the distance of the upstream mark to the downstream mark by the time taken by the floating object to travel the abovementioned distance, as shown in Eq. (6.3).

$$v = \frac{dS}{dt} \tag{6.3}$$

Now we need to measure the cross-sectional area of the open channel. This is done by determining the width of the channel and the depth of the channel. The width is measured by measuring the distance from one bank to the other bank on a straight line. The depth of the channel is measured by taking the measurements of the depth by an appropriate scale at multiple locations on the same length where the width was measured and average all the readings. The area is calculated by the following Eq. (6.4).

$$A = W \times d \tag{6.4}$$

Now the flow rate of the water stream is determined by multiplying the velocity of the water in the mainstream and the cross-sectional area of the stream channel, as illustrated in Eq. (6.5)

$$Q = v \times A \tag{6.5}$$

Flow duration curve (FDC) is a way of organizing the discharge data of a river or a stream in a graphical appearance that gives an estimation of the fraction of the time at which the flow rate equals or exceeds some value of interest. It is obtained by arranging the discharge rate chronologically and drawing it on the vertical axis at equal intervals of percentage on the horizontal axis. To well present the discharge behavior with higher accuracy, daily time-step measurements are needed to draw FDCs [1]. However, FDCs can be drawn for any specific time duration and any particular year [4].

The water flow at the site where the installation of the micro hydropower plant is intended is measured and recorded. The process is repeated daily and is averaged for the whole month. By the end of the year, you will have 12 readings of the flow rate. Now rearrange the data points in a table format in descending order from the highest value to the lowest one. Now we have to plot 12 data points on a percentage exceedance scale that will go from 0% to 100%. In our case, we will have $100\%/12 \cong 8.33$ percent increment on percentage exceedance scale. Table 6.3 shows the above-stated procedure to draw the FDC.

A FDC is drawn by using the data found in Table 6.3. As an example, a point is marked on the graph showing that 2 m^3/s water flow rate is available

TABLE 6.3 Flow rate and the % exceedance of the year.

Sr. no.	Month	Flow rate m^3/s	Percentage exceedance (%)
1	June	4.60	8.33
2	July	4.23	16.66
3	August	3.00	24.99
4	September	2.65	33.32
5	October	2.25	41.65
6	November	2.00	49.98
7	December	1.80	58.31
8	January	1.55	66.64
9	February	1.25	74.97
10	March	1.13	83.30
11	April	0.56	91.63
12	May	0.20	100

50% of the time throughout the year, as shown in Fig. 6.4. The FDC is useful while designing the micro hydropower system.

6.4.2 Weir and open channel

A weir is a structural barrier across the open channel streamline that is designed to change the flow characteristics of the water. Engineers and hydrologists use the weir to measure the volumetric flow rate in medium and small streams. Usually, it is a little wall or a minor dam with a notch to channel the water. Discharge can be calculated by merely measuring the level difference between the upper stream of water and the lower surface of the groove. Based on the material type and the shape, various kinds of grooves are used, such as trapezoidal, vee, and rectangular, made up of wood and/or metal plates. To measure the flow rate at the weir, its width must be greater than three times the height. Eq. (6.6) is used to measure the flow rate through the notch in a weir. Fig. 6.5 shows the dimensions of the weir.

$$Q = 1.8 \times (W - 0.2\,h) \times h^{\frac{3}{2}} \tag{6.6}$$

where W is weir width and h is weir height.

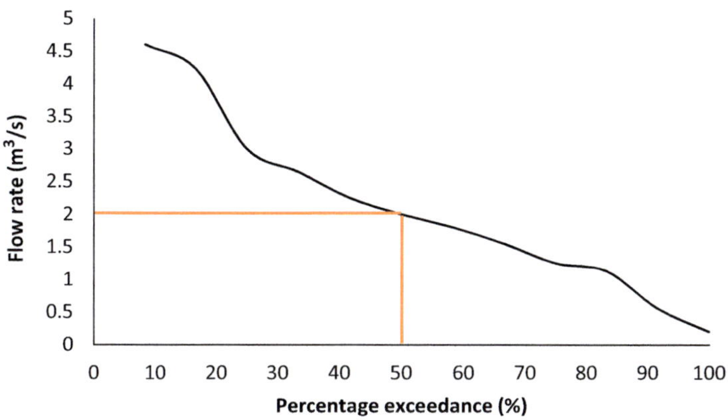

FIGURE 6.4 Flow duration curve.

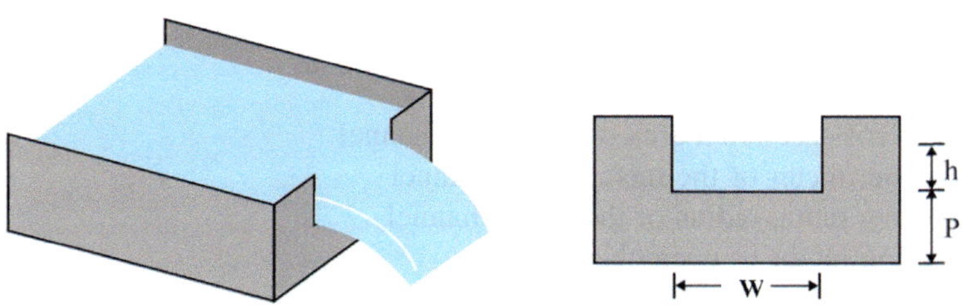

FIGURE 6.5 Dimensions of the weir.

In 1889 Robert Manning introduced an equation alternative to the Chezy equation that governs the open channel flow. The Manning equation is used in determining the open channel flow, capacity of a pipe, friction losses, and in various other applications. This equation is related to the uniform flow in the open channel and is shown in Eq. (6.7).

$$Q = \frac{1}{n}\left(sf \times S^{\frac{1}{2}}\right) \tag{6.7}$$

where n is Manning roughness coefficient, sf is section factor $= AR^{2/3}$, and S is channel bottom line slope.

Eq. (6.7) can be restructured to Eq. (6.8), which measures the bottom line slope of the channel.

$$S = \left(\frac{Q \times n}{A \times R^{\frac{2}{3}}}\right)^2 \tag{6.8}$$

To calculate the bottom line slope, we first need to calculate the open channel cross-sectional area and the hydraulic radius (R) of the cross-sectional area by using Eqs. (6.9) and (6.10), respectively.

$$A = W \times h \tag{6.9}$$

$$R = \frac{A}{P} = \frac{A}{W + 2\,h} \tag{6.10}$$

where A is open channel cross-sectional area and P is perimeter of the rectangular channel.

The open channel velocity is calculated by using Eq. (6.11) if the flow rate through the channel and the area of the channel are known.

$$v = \frac{Q}{A} \tag{6.11}$$

Example 6.3

It is intended to build a micro hydropower plant at a canal. The water is to channel to the penstock through the open channel and a rectangular weir 2 m wide, 1 m high, and with a 1.2 m perimeter. The open channel width and height are 5 and 2 m respectively. The bottom line slope of the open channel is 10. Calculate

1. The flow rate through the weir
2. The cross-sectional area of the open channel
3. The perimeter of the rectangular channel
4. The hydraulic radius of the open channel
5. Section factor of open channel
6. The flow rate through the open channel
7. Water velocity through the open channel

Solution
Given data

$$\text{Weir} \Rightarrow W = 2\ \text{m},\ h = 1\ \text{m},$$

$$\text{open channel} \Rightarrow W = 5\ \text{m},\ h = 2\ \text{m},\ S = 10$$

1. The flow rate through the weir is determined by Eq. (6.6)

$$Q = 1.8 \times (W - 0.2\,h) \times h^{\frac{3}{2}}$$

$$Q = 1.8 \times (2 - 0.2) \times (1)^{\frac{3}{2}}$$

$$Q = 3.24\ \frac{\text{m}^3}{\text{s}}$$

2. The cross-sectional area of the open channel is determined by Eq. (6.9)

$$A = W \times h$$

$$A = 5\ \text{m}\ \times\ 2\ \text{m}$$

$$A = 10\ \text{m}^2$$

3. The perimeter of the open channel is determined by the following equation

$$P = W + 2\,h \Rightarrow 5\ \text{m} + 2\ (2\ \text{m})$$

$$P = 9\ \text{m}$$

4. The hydraulic radius of the open channel is determined by Eq. (6.10)

$$R = \frac{A}{P} \Rightarrow \frac{10\ \text{m}^2}{9\ \text{m}}$$

$$R = 1.11\ \text{m}$$

5. Section factor of the open channel is determined by the following relation

$$sf = AR^{\frac{2}{3}}$$

$$sf = (10\ \text{m}^2)(1.11\ \text{m})^{\frac{2}{3}}$$

$$sf = 11.11$$

6. The flow rate through the open channel is determined by using the Manning equation given in Eq. (6.7).

$$Q = \frac{1}{n}\left(sf \times S^{\frac{1}{2}}\right)$$

$$Q = \frac{1}{0.012}\left(11.11 \times (10)^{\frac{1}{2}}\right)$$

$$Q = 2927.74\ \mathrm{m^3/s}$$

7. Water velocity through the open channel is determined using Eq. (6.11)

$$v = \frac{Q}{A} \Rightarrow \frac{2927.74\ \mathrm{m^3/s}}{10\ \mathrm{m^2}}$$

$$v = 292.77\ \mathrm{m/s}$$

6.4.3 Trash rack design

Trash racks are the obstructions to the flow to prevent the floating and immersed debris from getting into the intake and damaging the penstock and the turbine. Bars of some specific thickness with certain spacing are placed at an inclination of 60–80 degree with the horizontal axis. Nasir has reported some typical values of the bar spacing for different turbines (20–30 mm for Pelton, 40–50 mm for Francis, and 80–100 mm for Kaplan turbines). The debris accumulated in the trash rack needs to be removed regularly either manually or by some systematic mechanism, otherwise, they may cause some additional head losses.

6.4.4 Penstock design

The penstock is used to drain the water from the source to the hydro turbine in the powerhouse. This is the main part of the micro hydro as it converts the potential energy of the water into kinetic energy. It could be either exposed or buried under the ground based on various factors like the ground nature, penstock material, ambient temperature, and environmental concerns. While designing penstock, various parameters like diameter, thickness, water velocity, and cross-sectional area are considered.

6.4.4.1 Penstock diameter

The material of the penstock depends upon the water pressure, environmental concerns, and the local terrain conditions. The internal diameter of the penstock can be calculated by using Eq. (6.12).

$$D = 2.69 \times \left(n^2 \times Q^2 \times \frac{L}{H_g}\right)^{0.1875} \tag{6.12}$$

where D is internal diameter of the penstock, n is Manning roughness coefficient, Q is design flow, L is penstock length, and H_g is gross head.

TABLE 6.4 Manning coefficients of different surface materials.

Sr. no.	Surface material	Manning coefficient
1	Cement	0.011
2	Concrete	0.012
3	Clay tile	0.014
4	Glass	0.010
5	Plastic	0.009
6	Galvanized iron	0.016

The Manning coefficient is the hindrance that is faced by the water while flowing through the penstock or the water channels [5]. The Manning roughness coefficient depends upon the type of the material of the penstock. Manning coefficients of some of the materials are given in Table 6.4.

6.4.4.2 *Penstock thickness*

The wall thickness of the penstock depends upon the penstock diameter, operating pressure, penstock material, and tensile strength. If the diameter of the penstock is known, its thickness can be determined by Eq. (6.13), which gives the minimum thickness of the penstock.

$$t = \frac{D + 508}{400} + 1.2 \tag{6.13}$$

where t is minimum penstock wall thickness and D is internal diameter of the penstock.

6.4.4.3 *Penstock cross-sectional area*

To determine the area of the cross section of the penstock using Eq. (6.14), its radius or the diameter must be known.

$$A = \frac{\pi D^2}{4} = \pi r^2 \tag{6.14}$$

where A is cross-sectional area of the penstock, D is internal diameter of the penstock, and r is internal radius of the penstock.

6.4.4.4 *Water velocity through the penstock*

The water at the head of the penstock contains potential energy. As the water moves down the penstock, height is decreased and the velocity of the water is increased. Potential energy is lost and the kinetic energy of the falling water is gained [6].

$$\text{Loss in potential energy} = \text{Gain in kinetic energy}$$

$$mgh = \frac{1}{2}mv^2$$

$$v = \sqrt{2gh} \tag{6.15}$$

where m is mass of falling water, g is gravitational acceleration, h is height of the falling water, and v is velocity of the water.

Eq. (6.15) uses the height of the penstock at any point of interest. The velocity of the water in the penstock can also be determined by dividing the flow rate by the cross-sectional area of the penstock, as given in Eq. (6.11).

6.4.5 Penstock losses

6.4.5.1 Head friction losses

The power that is available in power at the head of the penstock is greater than the power that is available at the output of the electrical generator because of the losses. In calculating the actual power from the turbine, head or penstock losses cannot be overlooked. These are the losses occurring due to the water friction with the wall of the penstock and are caused by the reduction of the actual head. Head friction losses can be calculated by the Darcy–Weisbach formula given in Eq. (6.16).

$$h_f = f \times \left(\frac{L}{D}\right) \times \frac{v^2}{2g} \tag{6.16}$$

where f is friction factor, L is length of the penstock, D is diameter of the penstock, v is average velocity, and g is gravitational acceleration.

The friction factor (f), a dimensionless quantity, is a function of the pipe diameter, roughness coefficient, and the viscosity of the water and is calculated by Eq. (6.17).

$$f = 124.5 \times \frac{n^2}{D^{1/3}} \tag{6.17}$$

6.4.5.2 Inlet losses

When fluid enters a pipe from a large area to a small conduit, there comes variation in the speed of the fluid. In a micro hydropower system, when water enters from the open channel to the penstock, it experiences some resistance that is termed as the penstock losses. The entrance or inlet losses are determined by Eq. (6.18).

$$h_e = f_e \times \frac{v^2}{2g} \tag{6.18}$$

where h_e is the entrance losses and f_e is the entrance coefficient—conventionally 0.5 for micro hydro.

6.4.5.3 Valve losses

Eq. (6.19) is used to determine the valve losses in the penstock. The valve is used to control the flow of the water into the penstock. The values of the coefficient of the valve type depend upon the type of the valve.

$$h_v = f_v \times \frac{v^2}{2g} \tag{6.19}$$

where h_v is the valve losses and f_v is the coefficient of valve type.

Hence, the overall penstock losses can be determined by summing up all the above losses as given by Eq. (6.20).

$$\text{Head losses} = h_f + h_i + h_v$$

$$\text{Head losses} = \frac{v^2}{2g}\left(\frac{f \cdot L}{D} + f_e + f_v\right) \tag{6.20}$$

This equation gives the overall penstock losses as the water moves from the head to the turbine in the penstock. The head losses decrease the available head and for net head calculations, head losses are subtracted from the gross head as illustrated in Eq. (6.21).

$$\text{Net head} = \text{Gross head} - \text{Head losses} \tag{6.21}$$

Contrary to the gross head, the net head gives the actual hydraulic power available at the input of the turbine.

6.4.6 Hydraulic power

Micro hydro does not require any major civil construction, thus it has no environmental effects. Water from the running flow of the river/canal or natural fall is diverted and after electricity generation is sent back to the stream. The power capacity in the water mainly depends upon the flow rate and head of the falling water. Power is defined as the rate of change of energy with respect to time and mathematically can be written as Eq. (6.22).

$$P = \frac{dE}{dt} \tag{6.22}$$

A micro hydro uses the potential energy of falling water that is used to rotate the turbine coupled to the electrical generator. So Eq. (6.22) can be modified as Eq. (6.23).

$$P = \frac{d(mgh)}{dt} \tag{6.23}$$

Mass can be replaced with the product of density and volume ($m = \rho \times V$)

$$P = gh\frac{d(\rho V)}{dt}$$

$$P = \rho gh\frac{d(V)}{dt}$$

The time rate of change of volume is known as the flow rate (Q).

$$P_t = \rho g Q H_g \tag{6.24}$$

where P_t is theoretically available power, ρ is water density, g is gravitational acceleration, Q is flow rate, and H_g is gross head without deducting head losses.

Eq. (6.24) gives the theoretically available power in the water at the potential without incorporating the penstock losses.

6.4.7 Turbine power

After bringing the head losses and the system efficiency into account, the actual available power of the turbine can be calculated by using Eq. (6.25).

$$P_{act.} = \rho g Q H_n \eta \tag{6.25}$$

where H_n is net head after deducting head losses and η is system efficiency.

The system efficiency is the efficiency of the turbine and the electrical generator.

Example 6.4

It is intended to build a micro hydropower plant at a canal having a 10 m gross head and 10 m^3/second flow rate. The required length of the cemented penstock is 12 m. To hold the water in the penstock a butterfly valve is used. The combined efficiency of the turbine and the generator is 90%. Find

1. The diameter of the penstock
2. The thickness of the penstock
3. The cross-sectional area of the penstock
4. Water velocity through the penstock
5. Friction factor
6. Head friction losses
7. Inlet losses
8. Valve losses
9. Penstock losses
10. Net head
11. Theoretically available power in the water
12. Actual power at the output
13. If the turbine speed is 180 rpm, find the capacity factor of the turbine

Solution
Given data

$$H_g = 10 \text{ m}$$
$$L = 12 \text{ m}$$
$$n = 0.011$$
$$Q = 10 \text{ m}^3/\text{s}$$
$$f_e = 0.5$$
$$f_v = 0.1$$
$$\eta = 90\%$$
$$\rho = 1000 \text{ kg/m}^3$$
$$N = 180 \text{ rpm}$$

1. The diameter of the penstock is determined by Eq. (6.12).

$$D = 2.69 \times \left(n^2 \times Q^2 \times \frac{L}{H_g}\right)^{0.1875}$$

$$D = 2.69 \times \left((0.11)^2 \times (10)^2 \times \frac{12}{10}\right)^{0.1875}$$

$$D = 2.88 \text{ m}$$

2. The thickness of the penstock is determined by Eq. (6.13).

$$t = \frac{D + 508}{400} + 1.2$$

$$t = \frac{2.88 + 508}{400} + 1.2$$

$$t = 2.477 \text{ m}$$

3. The cross-sectional area of the penstock is determined by Eq. (6.14).

$$A = \frac{\pi D^2}{4}$$

$$A = \frac{\pi (2.88)^2}{4} = 6.5 \text{ m}^2$$

4. Water velocity through the penstock is calculated by Eq. (6.15).

$$v = \sqrt{2\,gh}$$

$$v = \sqrt{2\,g(10)} = 14.14 \text{ m/s}$$

5. The friction factor is determined by Eq. (6.17).

$$f = 124.5 \times \frac{n^2}{D^{\frac{1}{3}}}$$

$$f = 124.5 \times \frac{(0.011)^2}{(2.88)^{\frac{1}{3}}}$$

$$f = 0.010$$

6. Head friction losses are determined by Eq. (6.16).

$$h_f = f \times \left(\frac{L}{D}\right) \times \frac{v^2}{2g}$$

$$h_f = 0.010 \times \left(\frac{12}{2.88}\right) \times \frac{(14.14)^2}{2g}$$

$$h_f = 0.425 \text{ m}$$

7. Inlet losses of the penstock are calculated by Eq. (6.18).

$$h_e = f_e \times \frac{v^2}{2g}$$

$$h_e = 0.5 \times \frac{(14.14)^2}{2g}$$

$$h_e = 4.99 \text{ m}$$

8. Valve losses are determined by Eq. (6.19).

$$h_v = f_v \times \frac{v^2}{2g}$$

$$h_v = 0.1 \times \frac{(14.14)^2}{2g}$$

$$h_v = 0.999 \text{ m}$$

9. The penstock losses are determined by adding head friction losses, inlet losses, and the valve losses calculated in part (6), (7), and (8) respectively.

$$\text{Head losses} = h_f + h_i + h_v$$

$$\text{Head losses} = 0.425 \text{ m} + 4.99 \text{ m} + 0.999 \text{ m}$$

$$\text{Head losses} = 6.414 \text{ m}$$

10. The net head is determined by subtracting the head losses from the gross head.

$$\text{Net head} = \text{Gross head} - \text{Head losses}$$

$$\text{Net head} = 10\text{ m} - 6.414\text{ m}$$

$$H_n = 3.586\text{ m}$$

11. The power that is available in the water theoretically without incorporating losses and the system efficiency is determined by Eq. (6.24).

$$P_t = \rho g Q H_g$$

$$P_t = (1000)(9.8)(10)(10)$$

$$P_t = 980\text{ kW}$$

12. The actual power that we get at the output from the electrical generator is the power that is calculated by using the net head and using the system efficiency.

$$P_{act.} = \rho\, g\, Q\, H_n\, \eta$$

$$P_{act.} = (1000)(9.8)(10)(3.586)(90\%)$$

$$P_{act.} = 316\text{ kW}$$

13. The capacity factor of the turbine is determined by Eq. (6.26).

$$\sigma = \frac{2N\sqrt{\pi Q}}{(2gH_n)^{\frac{3}{4}}}$$

The value of n is given in rpm, in Eq. (6.26) it must be used in rps.

$$\sigma = \frac{2(180/60)\sqrt{\pi(10)}}{(2g \times 3.586)^{\frac{3}{4}}}$$

$$\sigma = 1.39$$

6.4.8 Turbine speed

The capacity factor of a micro hydro turbine is the fraction of the actual output power to the rated power that it can produce. Capacity factor (σ) is used to determine the type of turbine suitable for the hydro project under the study and if the turbine speed, flow rate, and the head are known, it can be determined by Eq. (6.26). On the other hand, if the head and the capacity factor are known, the type of turbine can be estimated by a chart, as shown in Fig. 6.6.

$$\sigma = \frac{2N\sqrt{\pi Q}}{(2gH_n)^{\frac{3}{4}}} \tag{6.26}$$

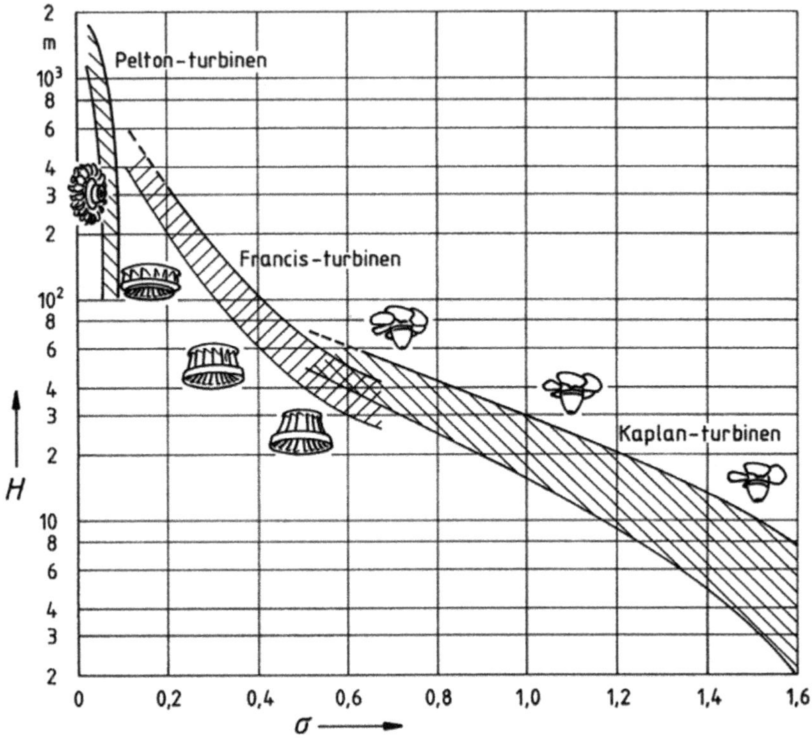

FIGURE 6.6 Application fields of water turbines with respect to head.

where σ is capacity factor of the turbine, N is rotational speed of the turbine, Q is designed flow rate, H_n is net head, and g is gravitational acceleration.

Eq. (6.27) is the rearranged form of Eq. (6.26) to find out the rotational speed of the turbine. The characteristic factor of the turbine is found from Fig. 6.6 at the net head of the system.

$$N = \frac{\sigma(2gH_n)^{\frac{3}{4}}}{2\sqrt{\pi Q}} \tag{6.27}$$

Example 6.5

A micro hydro turbine with a 15 m gross head is running at a speed of 180 rpm. The penstock losses are 2.5 m and the flow rate at the end of the penstock is 12 m^3/second. Calculate the capacity factor of the turbine.

Solution

We are given the gross head and the penstock losses. We first calculate the net head by subtracting the penstock losses from the gross head.

$$H_n = H_g - \text{losses}$$

$$H_n = 15\text{ m} - 2.5\text{ m}$$

$$H_n = 12.5\text{ m}$$

To calculate the capacity factor of the turbine we use Eq. (6.26).

$$\sigma = \frac{2N\sqrt{\pi Q}}{(2gH_n)^{\frac{3}{4}}}$$

$$\sigma = \frac{2\left(\frac{180}{60}\right)\sqrt{\pi(12)}}{(2 \times 9.8 \times 12.5)^{\frac{3}{4}}}$$

$$\sigma = 0.60$$

6.4.9 Specific speed

If a turbine is producing 1 kW power at a 1-m head, its speed will be known as the specific speed. The specific speed of the turbine is the point of maximum efficiency. The specific speed value for a turbine is the speed of a geometrically similar turbine which would produce unit power (1 kW) under unit head (1-m). The specific speed of the turbine is derived as follows.

From the power equation of the hydro turbine, we can deduct that the power is in direct relation to the flow rate and the head shown in Eq. (6.28).

$$P \propto QH \tag{6.28}$$

It is also well known to us that the absolute velocity of the turbine is directly proportional to the tangential velocity of the turbine, that is,

$$u \propto v \tag{6.29}$$

From Eqs. (6.15) and (6.29) we can deduct that the tangential velocity is in direct relation to the square root of the head shown in Eq. (6.30).

$$u \propto \sqrt{H} \tag{6.30}$$

From the tangential velocity formula $\left(u = \pi DN/60\right)$ and the equation we can deduct Eq. (6.31).

$$D \propto \frac{\sqrt{H}}{N} \tag{6.31}$$

As the flow rate is the product of the cross-sectional area and the water velocity, so from Eq. (6.11), we can deduct the following relation of Eq. (6.32).

$$Q \propto Av \tag{6.32}$$

By using Eqs. (6.14) and (6.30) we can write

$$Q \propto D^2\sqrt{H} \tag{6.33}$$

By Eqs. (6.28), (6.31), and (6.33) we can further have the following relation of Eq. (6.34)

$$P = K\frac{H^{\frac{5}{2}}}{N^2} \tag{6.34}$$

By the definition of the specific speed, power must be 1 kW at 1 m head, so in Eq. (6.34), $N = N_s$, which gives us the value of the proportionality constant K.

$$K = N_s^2 \tag{6.35}$$

So after putting the value of K from Eq. (6.35) and rearranging Eq. (6.34), we have Eq. (6.36) for the specific speed.

$$N_s = \frac{N\sqrt{P}}{H^{\frac{5}{4}}} \tag{6.36}$$

By the specific speed of the turbine, we can decide which type of hydro turbine will be used for the specific site of the project. Table 6.5 shows the relation between the type of the turbine and the specific speed.

Example 6.6

A 90% efficient turbine of the micro hydro system running at 150 rpm develops a power of 15 kW at net head 10 m. Calculate the specific speed of the turbine and the type of turbine.

Solution

The specific speed of the turbine is determined by using Eq. (6.36).

$$N_s = \frac{N\sqrt{P}}{H^{\frac{5}{4}}}$$

$$N_s = \frac{(150\text{ rpm})\sqrt{15\text{ W}}}{(10\text{ m})^{\frac{5}{4}}}$$

$$N_s = 32.66\text{ rpm}$$

The turbine for this specific speed will be a Pelton wheel with two or more jets as shown in Table 6.5.

TABLE 6.5 Type of turbine with respect to the specific speed.

Sr. no.	Type of turbine	Specific speed (N_s) (rpm)
1	Pelton wheen with single jet	8.5–30
2	Pelton wheel with two or more jets	30–51
3	Francis turbine	51–225
4	Kaplan turbine or propeller	225–860

6.4.10 Turbine selection

Once the net head and the capacity factor are found, the type of turbine suitable for the naturally occurring hydro site can be estimated from Fig. 6.6. Only the Pelton turbines are suitable for the large heads and low flow. Francis turbines can be used in the middle area from around 500 m. In the area where the head overlaps with the Pelton turbines of around 100–500 m, the Pelton turbine is more insensitive to wear, but the Francis turbine is superior due to its higher efficiency, better application possibilities with fluctuating heads, and smaller dimensions. Kaplan turbines are used for the smaller heads. Here, too, there is an overlap with the Francis turbines in the range of around 40–50 m, where the Kaplan turbines are usually preferred. Fig. 6.6 shows the chart of the turbines to be used at different heads.

Once the type of turbine is identified, other primary dimensions of the turbine can be found. The inner diameter of the hub, the outer diameter of the wheel, and the ratio D_N/D_a can be found by using Eqs. (6.37), (6.38), and (6.39), respectively.

$$D_a = \frac{2\delta}{\pi} \times \sqrt[4]{\frac{Q^2}{2gH}} \tag{6.37}$$

$$D_N = 4D_a \tag{6.38}$$

$$\text{diameters ratio} = \frac{D_N}{D_a} \tag{6.39}$$

where D_a is hub inner diameter, δ is diameter number, and D_N is the outer diameter of the turbine.

If the characteristic factor (σ) is marked in the Kaplan turbine region in the chart shown in Fig. 6.7, the inner diameter of the hub, the outer diameter of the wheel, diameter number (δ), and the ratio D_N/D_a can be identified as shown in Fig. 6.7.

Example 6.7

A turbine has to be selected for the micro hydropower system with a 15 m head, 8 m/second flow rate, and 1.8 diameter number. Calculate

1. The hub inner diameter
2. The outer diameter of the turbine
3. The diameter ratio

Solution

1. The inner diameter of the hub is determined by Eq. (6.37)

$$D_a = \frac{2\delta}{\pi} \times \sqrt[4]{\frac{Q^2}{2gH}}$$

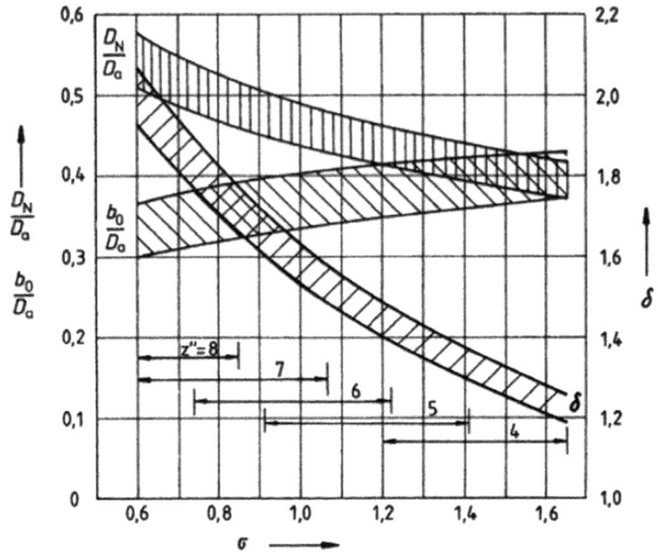

FIGURE 6.7 Design diagram for a hydro turbine.

$$D_a = \frac{2(1.8)}{\pi} \times \sqrt[4]{\frac{(8)^2}{2g(15)}}$$

$$D_a = 1.04 \text{ m}$$

2. The outer diameter of the turbine is determined by Eq. (6.38)

$$D_N = 4D_a \Rightarrow 4(1.04)$$

$$D_N = 4.16 \text{ m}$$

3. The diameter ratio is determined as follow.

$$\text{Diameters ratio} = \frac{D_N}{D_a} \Rightarrow \frac{4.16 \text{ m}}{1.04 \text{ m}}$$

$$\text{Diameters ratio} = 4.0$$

6.5 SWOT analysis of hydropower

6.5.1 Strengths

- Hydropower is a renewable energy source that can replenish itself after completing its life cycle.

- It's an environment-friendly source of energy as it does not emit any health hazard emissions, such as NO_x and SO_x. It does not have any share in global warming as it does not emit carbon dioxide.
- Long lifetime of the project. Approximately more than 50 years.
- Lower cost of energy, operation, and maintenance as compared to other renewable energy sources.
- Higher efficiency of hydropower as compared to conventional and non-conventional energy sources.

6.5.2 Weaknesses

- Usually, the hydro potential is available in remote hilly areas due to available water resources at the elevated location. This increases the transmission cost and transmission losses.
- In remote areas, construction of the storage dam may delay the commercial operation date of the project because of the unavailability of the construction materials at the locality.
- Deforestation may harm the environment.
- Degradation of the existing dams and hydropower plants by sedimentation.

6.5.3 Opportunities

- The available hydro potential is still unharnessed.
- Other potential activities associated with hydro like flood control, irrigation, tourism, potable water supply, recreation, and navigation, etc.
- Development of remote and backward areas.
- Adding capacities to the already existing dams.
- Unexploited potential allows harnessing the potential that is still untapped.
- Flood risk management can control the flood in the rainy and flooded season.

6.5.4 Threats

- Large dams are considered to be a heavily concentrated load on the ground that may cause the earthquake.
- Restoration of people and land acquisition for construction of the dam and electrical facility.
- Due to heavy rains, the safety of the dams becomes crucial as the heavy discharge from the dam may cause a flood in the downstream areas.
- Demolition of fish and wildlife habitat.

Problems

Problems 1–9 contain three/four answer options A, B, C, and D. Choose the correct answer.

1. From the given methods, which one is used in hydropower?
   A. Potential energy of water is converted into mechanical energy of turbine
   B. Kinetic energy of water is converted into mechanical energy of turbine
   C. Heat energy of water is converted into mechanical energy of turbine
   D. Ionization of water is used to rotate the turbine
2. Which of the following is a type of impulse turbine?
   A. Francis
   B. Propeller
   C. Pelton
   D. Pump as a turbine
3. Which of the following is a type of reaction turbine?
   A. Francis
   B. Pelton
   C. Turgo
   D. Cross-flow
4. Reaction turbines are used at
   A. Low head
   B. Medium head
   C. High head
   D. A & B
5. In a hydro plant, if the discharge is 200 m^3/second and the head of the water is 100 m. If the efficiency of the turbine alternator is set to 0.85, find the power developed
   A. 177 MW
   B. 166.6 MW
   C. 66.67 MW
   D. 150.7 MW
6. To draw the FDC, discharge data is first arranged in
   A. Ascending order
   B. Descending order
   C. Apple pie order
   D. First-in first-out order
7. In FDC data is arranged in ---------------- against the flow rate
   A. Percentage exceedance
   B. Incremental order
   C. Decremental order
   D. Ascending order
8. Which is the correct formula for the bottom line slope
   A. $S = \left(\frac{Q \times h}{A \times R^{\frac{2}{3}}}\right)^2$
   B. $S = \left(\frac{Q \times n}{A \times R^{\frac{2}{3}}}\right)^2$
   C. $S = \left(\frac{Q \times P}{A \times R^{\frac{2}{3}}}\right)^2$

9. The diameter of the penstock is 5 m. What will be the thickness of the penstock?
   A. 2.48 m
   B. 3.01 m
   C. 2.01 m
   D. 2.60 m
10. What is the difference between an impulse turbine and reaction turbine? Write down three types of each turbine.
11. How are FDC s drawn for a specific site of the hydro project?
12. Hydraulic radius of an open channel is 2 m and the cross-sectional area is 12.6 m^2. Calculate the section factor.
13. Write down the difference between gross head and net head.
14. What are the penstock losses? How they are helpful in calculating the net head?
15. The flow rate at the 25 m gross head is 10 m^3/second with 4 m penstock losses. If the system is 90% efficient, calculate the output power of the turbine.
16. Give a brief SWOT analysis of the hydropower.

References

[1] M. Pérez-Sánchez, F.J. Sánchez-Romero, H.M. Ramos, P.A. López-Jiménez, Energy recovery in existing water networks: towards greater sustainability, Water 9 (2) (2017) 1–20. Available from: https://doi.org/10.3390/w9020097.

[2] N. Belyakov, Chapter Fourteen—traditional hydropower plant technology, in: N.B.T.-S.P. G. Belyakov (Ed.), Sustainable Power Generation, Academic Press, 2019, pp. 355–377.

[3] S. Lajqi, N. Lajqi, B. Hamidi, Design and construction of mini hydropower plant with propeller turbine, Int. J. Contemp. Energy 2 (1) (2016) 1–13. Available from: https://doi.org/10.14621/ce.20160101.

[4] B.A. Nasir, Design considerations of micro-hydro-electric power plant, Energy Proc 50 (2014) 19–29. Available from: https://doi.org/10.1016/j.egypro.2014.06.003.

[5] M. Kamran, M.R. Fazal, M. Mudassar, Towards empowerment of the renewable energy sector in Pakistan for sustainable energy evolution: SWOT analysis, Renew. Energy 146 (2020), Available from: https://doi.org/10.1016/j.renene.2019.06.165.

[6] M. Kamran, R. Asghar, M. Mudassar, M.I. Abid, Designing and economic aspects of run-of-canal based micro-hydro system on Balloki-Sulaimanki Link Canal-I for remote villages in Punjab, Pakistan, Renew. Energy 141 (2019). Available from: https://doi.org/10.1016/j.renene.2019.03.126.

Chapter 7

Fuel cell

Muhammad Kamran
Department of Electrical Engineering and Technology, Riphah International University, Pakistan

7.1 Introduction

Fuel cells (FCs) are the electrochemical devices that convert the chemical energy of a reaction into DC electrical energy. In most FCs, hydrogen is used as a source of chemical energy that is oxidized at the anode splitting hydrogen into electrons and protons. The protons move through the proton exchange membrane and reach the cathode. The electrons move through the load in the external circuit and reach the cathode. At the cathode, reduction of the oxygen takes place combining the oxygen, proton, and electron-generating water as a by-product. As no moving parts are involved and no combustion occurs in the FC, no wear and tear maintenance is required and no NOx is generated. As the sulfur contaminates the FC, so all the sulfur content must be removed from the fuel before using it in the FC, hence no SOx is produced. A FC can be brought into operation on instant demand as long as the fuel and the oxygen are present.

7.2 Working principle of a fuel cell

FCs generate DC electricity by an electrochemical reaction. FCs consist of two electrodes, named the anode and cathode, the cathode gas diffusion layer, the anode gas diffusion layer, the cathode catalyst layer, the anode catalyst layer, and a proton exchange membrane of polymer that separates the anode and the cathode portions. To understand the basic working principle of a FC, we discuss the hydrogen-based proton exchange membrane FC (PEMFC). The detailed structure of a PEMFC is shown in Fig. 7.1. From the hydrogen storage tank, hydrogen gas enters the FC from the anode side. Porous and thin gas diffusion layers at the anode and cathode provide a suitable flow for the reactant gases into the catalyst layer. It also provides corrosion resistance and thermal and electrical conductivity. A catalyst of platinum mixed with an ion-conducting polymer ionomer is sandwiched between the anode gas diffusion layer and the membrane that helps in splitting the incoming hydrogen molecules into electrons and protons. The protons diffuse over the proton exchange membrane to the cathode gas diffusion layer, whereas the electrons move toward the cathode

Renewable Energy Conversion Systems. DOI: https://doi.org/10.1016/B978-0-12-823538-6.00005-1

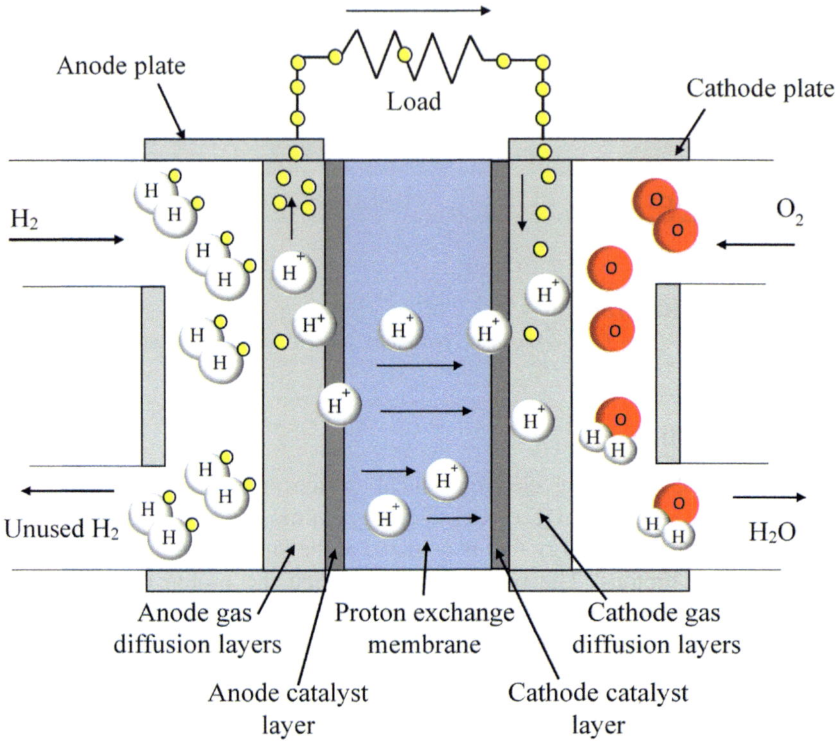

FIGURE 7.1 Proton exchange membrane fuel cell.

through the anode and the external DC load. Platinum on the cathode catalyst layer helps the reduction of oxygen molecules into two oxygen atoms where they attract two protons coming through the Proton Exchange Membrane (PEM) and two electrons coming from the cathode to produce water.

The oxidation reaction of hydrogen taking place at the anode is shown in Eq. (7.1), where the hydrogen molecules are split into two protons and two electrons. Eq. (7.2) shows the cathode reaction combing the oxygen, proton, and electrons to make water. Eq. (7.3) shows the overall reaction occurring in the FC.

$$\text{Anode: } 2H_2 \longrightarrow 4H^+ + 4e^- \tag{7.1}$$

$$\text{Cathode: } O_2 + 4H^+ + 4e^- \longrightarrow 2H_2O \tag{7.2}$$

$$\text{Overall: } 2H_2 + O_2 \longrightarrow 2H_2O \tag{7.3}$$

7.3 Maximum efficiency of a fuel cell

7.3.1 Enthalpy of a reaction

Enthalpy of a chemical reaction, also known as the heat of a chemical reaction, is defined as the sum of the change in internal energy during the chemical reaction plus the pressure–volume work shown in Eq. (7.4).

$$\Delta H = \Delta U + P\Delta V \tag{7.4}$$

The first law of thermodynamics states that a part of the heat added to a system is used to do some external work while the rest is used to raise the internal energy of the system. So, by the first law of thermodynamics, change in internal energy is the heat added to the system plus the work done by the system. If the volume of the system increases during a chemical reaction, the work done by the system (PΔV) becomes negative. So Eq. (7.4) takes the form of Eq. (7.5)

$$\Delta H = q + W - W \tag{7.5}$$

$$\Delta H = q \tag{7.6}$$

By Eq. (7.6) we can interpret the type of reaction, that is, whether it is an exothermic reaction or an endothermic reaction. At constant temperature and pressure, the change in enthalpy of the reaction will be negative if the reaction takes heat from the surroundings. Similarly, the change in enthalpy of the reaction will be positive if the reaction releases heat to the surroundings. The enthalpy change for a chemical reaction is determined by using Eq. (7.7) of the chemical reaction.

Change in enthalpy (ΔH) of a chemical reaction is found by deducting the enthalpy of formation of reactants from the enthalpy of formation of the products using Eq. (7.8).

$$\Delta H = \sum (h_f)_{products} - \sum (h_f)_{reactants} \tag{7.7}$$

$$\Delta H = (c.h_f)_{product1} + (d.h_f)_{product2} - (a.h_f)_{reactant1} - (b.h_f)_{reactant2} \tag{7.8}$$

where
ΔH is the change in enthalpy
a is the coefficient of reactant 1 from the balanced chemical equation
b is the coefficient of reactant 2 from the balanced chemical equation
c is the coefficient of product 1 from the balanced chemical equation
d is the coefficient of product 2 from the balanced chemical equation
h_f is the enthalpy of formation

Change in enthalpy of the formation of water given in Eq. (7.8) of the chemical reaction can be determined using the above-stated procedure. Eq. (7.7) can be redrawn as Eq. (7.9) where water is the product, hydrogen, and oxygen are the reactants.

$$\Delta H = (h_f)_{H_2O} - (h_f)_{H_2} - \frac{1}{2}(h_f)_{O_2} \tag{7.9}$$

$$\Delta H = -286\ kJ/mol$$

TABLE 7.1 Entropy and enthalpy of formation of different elements.

	h_f (kJ/mol)	S_f (kJ/mol)
Water, H_2O (g)	−241.98	0.18384
Water, H_2O (L)	−286.02	0.06996
Oxygen, O_2 (L)	0	0.20517
Hydrogen, H_2 (L)	0	0.13066

Enthalpy of formation of water, hydrogen, and oxygen can be taken from Table 7.1 and put in Eq. (7.9). The negative value of the ΔH indicates that it is an endothermic reaction and heat is released during the formation of water.

7.3.2 The entropy of a reaction

In nature, every system tends to be more disordered. The measure of randomness or disorder is called the entropy of the system. Solids have less entropy as the structure of solids is more ordered. Whereas liquids and gases have more entropy as their structures are more disordered. If there is a transition from a gas to liquid or solid during a chemical reaction, the total entropy of the system is decreased as the system has moved from a disordered structure to a more ordered structure.

Change in entropy (ΔS) of a chemical reaction is found by deducting the entropy of formation of reactants from the entropy of formation of the products using Eq. (7.10).

$$\Delta S = \sum (S_f)_{products} - \sum (S_f)_{reactants} \tag{7.10}$$

$$\Delta S = (c \cdot S_f)_{product1} + (d \cdot S_f)_{product2} - (a \cdot S_f)_{reactant1} - (b \cdot S_f)_{reactant2} \tag{7.11}$$

where
ΔS is the change in entropy
a is the coefficient of reactant 1 from the balanced chemical equation
b is the coefficient of reactant 2 from the balanced chemical equation
c is the coefficient of product 1 from the balanced chemical equation
d is the coefficient of product 2 from the balanced chemical equation
S_f is the entropy of formation

Change in entropy of formation of water given in Eqs. (7.10) and (7.11) of a chemical reaction can be determined using the above-stated procedure. Eq. (7.10) can be redrawn for Eq. (7.12) where water is the product, hydrogen, and oxygen are the reactants.

$$\Delta S = (S_f)_{H_2O} - (S_f)_{H_2} - \frac{1}{2}(S_f)_{O_2} \tag{7.12}$$

$$\Delta S = 0.06996 - 0.13066 - \frac{1}{2}(0.20517)$$

$$\Delta S = -0.163 \text{ kJ/mol}$$

The entropy of the formation of water, hydrogen, and oxygen can be taken from Table 7.1 and put in Eq. (7.12). The negative value of the ΔS indicates that it is an endothermic reaction and heat is released during the formation of water.

7.3.3 Gibbs free energy

Gibbs's free energy is associated with the energy that is available to do work and it determines whether a reaction will be spontaneous or not. Like the universe, every system wants to be stable by giving off energy in which case the system will have a negative enthalpy change. On the other hand, the universe also wants to be disordered, that is, increase in entropy. Fulfilling these two conditions, the change in free energy will be negative and the reactions will be spontaneous on all values of temperatures. If the ΔG is positive, it means the reaction needs external energy to proceed and the reaction will not be a spontaneous one. Mathematically, Gibbs free energy is determined by subtracting the product of temperature and entropy change from the enthalpy change, as presented in Eq. (7.13).

$$\Delta G = \Delta H - T \cdot \Delta S \tag{7.13}$$

$$\Delta G = -286 \text{ kJ/mol} - 298(-0.163 \text{ kJ/mol})$$

The spontaneity of the reaction can be determined by looking at the ΔG and ΔH. Fig. 7.2 shows the different scenarios for checking the spontaneity of the reaction.

At standard temperature and pressure, the values of the change in enthalpy and the change in entropy during the formation of water calculated above are used here to determine the Gibbs's free energy available during the formation of water. On locating the value of ΔS and ΔH in Fig. 7.2, we can deduct that the change in free energy ΔG of the formation of water is negative leading the reaction toward spontaneity.

$$\Delta G = -237 \text{ kJ/mol}$$

7.3.4 The efficiency of a fuel cell

The thermodynamic efficiency of a FC is given by the ratio of the change in Gibbs free energy to the enthalpy change of the system. The efficiency of a FC using hydrogen as a fuel can be calculated using the values of ΔG and

	$\Delta H < 0$	$\Delta H > 0$
$\Delta S > 0$	$\Delta G < 0$ System is spontaneous at all temperatures	System is spontaneous at high temperatures, when $T\Delta S$ is large
$\Delta S < 0$	System is spontaneous at low temperatures, when $T\Delta S$ is small	$\Delta G > 0$ System is non spontaneous at all temperatures

FIGURE 7.2 Different scenarios for spontaneity of the reaction.

ΔH determined in section 7.3. Eq. 7.15 gives the maximum efficiency of the fuel cell.

$$\eta = \frac{\text{Available output energy}}{\text{Input energy from hydrogen}} \tag{7.14}$$

$$\eta = \frac{\Delta G}{\Delta H} \Rightarrow \frac{-237\ \text{kJ/mol}}{-286\ \text{kJ/mol}} \qquad \eta = 83\% \tag{7.15}$$

7.4 Fuel cell potential

The open-circuit voltage of the FC is defined as the potential across the electrodes of the cell when no current is flowing through the external connections. Conventionally, it is measured at a specific given temperature by taking the potential difference of the anode and cathode potential. Mathematically it can be written as

$$E_T = E_{T,C} - E_{T,A}$$

where

E_T is the open-circuit voltage at temperature T

$E_{T,C}$ is the potential drop at the cathode at temperature T

$E_{T,A}$ is the potential drop at the anode at temperature T

The potential of the reaction occurring at the anode and the cathode is determined by applying the Nernst equation at each electrode. Then the open-circuit voltage of the FC is determined by using Eq. (7.16).

$$E_T = \dot{E}_T + \left(\frac{RT}{nF}\right) \ln\left(\frac{(P_{\text{reactants}})^x}{(P_{\text{products}})^y}\right) \tag{7.16}$$

where

E_T is the open-circuit voltage of the FC at temperature T

$\dot{E}_T$is the standard potential of the reaction

R is the general gas constant

T is the operating temperature

n is the number of electrons involved

F is the Faraday's number, 96487 C/mol

$P_{\text{react.}}$ is the partial pressure of the reactant

$P_{\text{prod.}}$ is the partial pressure of the products

x, y are the coefficient of reactant, and products respectively from the balanced chemical equation

Eq. 7.16 is the general Nernst equation that can be used to determine the open-circuit voltage of any type of FC. We use the above-stated approach to determine the open-circuit voltage relation for the PEMFC discussed in Section 7.2. The chemical reaction occurring at the anode and cathode are shown in Eqs. (7.2) and (7.3).

7.4.1 At anode

Using the Nernst equation for the reaction occurring at the anode we get the potential at the anode of the PEMFC using Eq. (7.17). The chemical reaction at the cathode is shown in Eq. (7.3).

$$E_A = \dot{E}_A + \left(\frac{RT}{4F}\right) \ln\left(\frac{(P_{H^+})^4}{(P_{H_2})^2}\right) \tag{7.17}$$

$$E_A = E_A + \left(\frac{RT}{4F}\right) \left[4\, ln\, (P_{H^+}) - 2 \ln\, (P_{H_2})\right] \tag{7.18}$$

7.4.2 At cathode

Similarly, we use Eq. (7.16) to determine the relation of the open-circuit voltage at the cathode of the PEMFC using Eq. (7.19) and Eq. (7.20). The chemical reaction at the cathode is shown in Eq. (7.2)

$$E_C = \dot{E}_C + \left(\frac{RT}{4F}\right) \ln\left(\frac{(P_{O_2})^1 (P_{H^+})^4}{(P_{H_2O})^2}\right) \tag{7.19}$$

$$E_C = \dot{E}_C + \left(\frac{RT}{4F}\right)\left[\ln\left(P_{O_2}\right) + 4\ln\left(P_{H^+}\right) - 2\ln\left(P_{H_2O}\right)\right] \tag{7.20}$$

Now we sum up the Nernst equations of anode and cathode to determine the open-circuit voltage relation for the PEMFC.

$$E = \left[\dot{E}_C + \left(\frac{RT}{4F}\right)\left[\ln\left(P_{O_2}\right) + 4\ln\left(P_{H^+}\right) - 2\ln\left(P_{H_2O}\right)\right]\right] - \left[\dot{E}_A + \left(\frac{RT}{4F}\right)\left[4\,ln\left(P_{H^+}\right) - 2\ln\left(P_{H_2}\right)\right]\right]$$

On simplification of the above relation, we get the general Nernst equation for the PEMFC.

$$E = \dot{E} + \left(\frac{RT}{4F}\right)\left[2\ln\left(P_{H_2}\right) + \ln\left(P_{O_2}\right) - 2\ln(P_{H_2O})\right] \tag{7.21}$$

Eq. (7.21) is used to determine the open-circuit voltage of the PEMFC.

7.5 Terminal voltage of the fuel cell

The voltage that is available at the output of the FC when a load is connected will always be less than the internally generated voltage or the open-circuit voltages because of the losses encountered by the current as it moves to the external circuit. The losses that are faced by the current are categorized as activation loss, impedance loss, and concentration loss. These losses are subtracted from the open-circuit voltage to get the terminal voltages of the FC using Eq. (7.22). If N cells are connected in a stack, the terminal voltage of the stack of the fuel cells is calculated by Eq. (7.23)

$$V_T = E_T - \eta_{act} - \eta_{con} - \eta_{ohmic} \tag{7.22}$$

$$V_T = E_T - L$$

$$V_T = N(E_T - L) \tag{7.23}$$

where
V_T is the terminal voltage of the FC
E_T is the open-circuit voltage of the FC
η_{act} is the activation losses
η_{con} is the concentration losses
η_{ohmic} is the ohmic losses
L is the voltage losses $= \eta_{act} + \eta_{con} + \eta_{ohmic}$
N is the no. of FCs in the stack

7.5.1 Activation losses

When the electrochemical reaction takes place in the FC and the current starts flowing, the chemical reaction slows down because of the kinetics of the reaction at the anode and cathode. For example, in a direct methanol FC (DMFC), the slow oxidation of the methanol at the cathode and the slow reduction of the oxygen at the cathode of the PAFC create activation losses. The activation losses are determined by Eq. (7.24)

$$\eta_{act} = \frac{RT}{anF}\ln\left(\frac{I}{I_0}\right) \tag{7.24}$$

where

a is the electron transfer coefficient or load transfer coefficient, 0.5 value is assumed for hydrogen electrode, 0.1–0.5 for oxygen electrode

I is the load current

I_0 is the exchange current

Another type of activation losses is the fuel crossover and internal current. When the hydrogen atoms break into two electrons and two protons, the electrons do not move to the external circuit. Instead, they pass through the electrolyte and recombine the proton and oxygen on the cathode side. Hence these electrons do not participate in the current flowing through the external circuit. This type of current is termed as the internal current and is incorporated in the activation losses by Eq. (7.25).

$$\eta_{act} = \frac{RT}{anF}\ln\left(\frac{I+I_n}{I_0}\right) \tag{7.25}$$

where

I_n is the internal current

Activation losses can be minimized by the following methods.

- By increasing the temperature of the FC.
- By using the more effective catalyst which in return increases the activity of the electrodes.
- By increasing the concentration of the reactants by using pure oxygen instead of air.

7.5.2 Concentration losses

During the chemical reaction in the FC reactants must continuously be supplied and the products must continuously be removed from the FC since these concentrations determine the performance of the FC. The concentration of the reactants ensures the production of electricity and the concentration of the products determines the efficiency of the FC. This means that the gases near the electrolyte are consumed very quickly and may not be replaced as quickly as needed, thus decreasing the concentration around the catalyst. Concentration losses also known as mass transportation losses are determined by Eq. (7.26)

$$\eta_{con} = \frac{RT}{nF}\ln\left(\frac{I_{lim}}{I_{lim}-1}\right) \tag{7.26}$$

where

I_{lim} is the maximum current drawing

7.5.3 Ohmic losses

The hindrance to the flow of electrons by electrodes, gas diffusion layer, catalyst layer, the electrolyte, and the external connections are termed as the ohmic losses. Ohmic losses are calculated by Ohm's law, as stated in Eq. (7.27)

$$\eta_{ohmic} = iR_{ohmic} \tag{7.27}$$

Ohmic losses can be minimized by

- Using the highly conductive electrodes.
- Good design of the bipolar metal plates and their interconnections.
- Using thin electrolytes to give mechanical support to the flow of electrons.

Hence, Eq. (7.28) can be rewritten to determine the terminal potential of the FC.

$$V_T = E_T - \frac{RT}{anF}\ln\left(\frac{I}{I_0}\right) - \frac{RT}{nF}\ln\left(\frac{I_{lim}}{I_{lim}-1}\right) - \eta_{ohmic} \tag{7.28}$$

7.6 Equivalent circuit model of the fuel cell

The equivalent circuit model of the FC is determined by using the terminal voltage, FC, activation losses, concentration losses, and ohmic losses. In the equivalent circuit model of the FC, activation voltage losses and the concentration voltage losses are represented by the current controlled voltage sources, whereas the Nernst factor is represented by the voltage-controlled voltage source and is given in Eq. (7.29)

$$V_f = \left(\frac{RT}{nF}\right)\ln\left(\frac{(P_{\text{reactants}})^x}{(P_{\text{products}})^y}\right) \tag{7.29}$$

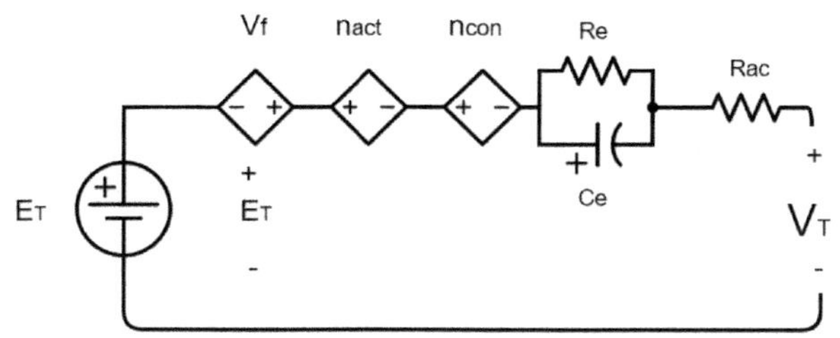

7.7 Types of fuel cell

7.7.1 Direct methanol fuel cell

DMFCs directly convert the chemical energy of the methanol–water into electrical energy. DMFCs consist of two electrodes working as anode and cathode, anode and cathode gas diffusion layers, anode and cathode catalyst layers of platinum, and a proton exchange membrane every so often made up of polymer that allows protons to pass through and keeps the anode and the cathode apart. The structure of the DMFC is shown in Fig. 7.3. Methanol plus water is used as a fuel that enters the FC from the anode side where it is oxidized at the anode platinum catalyst layer to form carbon dioxide. The reaction occurring at the anode is shown in Eq. (7.30). Methanol is oxidized into CO_2, six protons, and six electrons. CO_2 is drained to the outlet at the anode side, whereas the liberated electrons move toward the anode plate to reach the cathode through the external electrical load. The protons move through the proton exchange membrane to reach the cathode gas diffusion layer, where they attract the electrons that were generated during the methanol oxidization and the oxygen coming from the inlet of the FC at the cathode side to form water. Water and heat as a by-product in this reaction are collected at the outside of the FC. The reaction occurring at the cathode is shown in Eq. (7.31). Eq. (7.32) shows the overall reaction occurring in DMFC.

$$\text{Anode}: CH_3OH + H_2O \longrightarrow 6H^+ + 6e^- + CO_2 \tag{7.30}$$

$$\text{Cathode}: \frac{3}{2}O_2 + 6H^+ + 6e^- \longrightarrow 3H_2O \tag{7.31}$$

$$\text{Overall}: CH_3OH + \frac{3}{2}O_2 \longrightarrow 2H_2O + \frac{1}{2}CO_2 \tag{7.32}$$

Oxidation of methanol–water is a much more complex process as compared to the oxidation of hydrogen. The oxidation of methanol is completed in various steps that make the anode process much slower than the reduction of oxygen at the cathode. The complexity carried by the oxidation of methanol is detailed in Eqs. (7.33) and (7.34). The formation and later absorption of CO and COH during the methanol oxidation resist the methanol oxidation

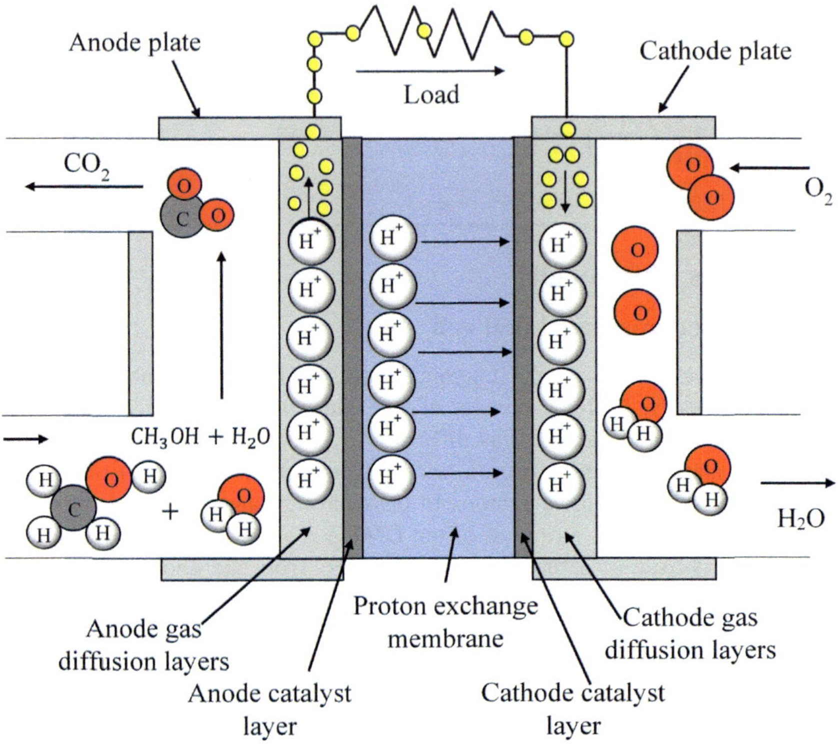

FIGURE 7.3 Direct methanol fuel cell.

on the Pt-catalyst. To alleviate the absorption of intermediate products on the anode catalyst layer, Pt is alloyed with any of the secondary metals like Ni, Co, Mo, Sn, and Ru.

$$CH_3OH + Pt \longrightarrow Pt - CH_2OH + H^+ + e^- \tag{7.33}$$

$$Pt - CH_2OH \longrightarrow Pt - CHOH + H^+ + e^-$$

$$Pt - CHOH \longrightarrow Pt - COH + H^+ + e^-$$

$$Pt - COH \longrightarrow Pt - CO + H^+ + e^-$$

$$Pt - CO + M - H_2O \longrightarrow Pt - M + CO_2 + 2H^+ + 2e^- \tag{7.34}$$

As a result of the slow oxidation of methanol, it passes through the anode, dissolves in the proton exchange membrane, and reaches the cathode catalyst layer where it creates hindrance in the reduction of the oxygen. This degrades the performance of the FC. Instead of using pure Pt at the cathode catalyst layer, an alloy of platinum with metals (Pt-M, where M could be any of Cr, Pd, Au, Fe) is preferred to mitigate the methanol crossover impact. DMFC has various advantages over other types of FCs.

Example 7.1: The reaction of the methanol FC is shown in Eq. (7.32). the enthalpy of formation and the Gibbs free energy of the reactants and the products are shown in the table. You are required to find out:

1. The change in Gibbs free energy that can be converted into useful work.
2. Change in enthalpy of the reaction
3. The efficiency of the conversion of free energy into useful work

	h_f (kJ/mol)	G_f (kJ/mol)
Carbon dioxide, CO_2	−393.5	−394.36
Water, H_2O	−286.02	−237.13
Oxygen, O_2	0	0
Methanol, CH_3OH	−239	−166.27

Solution

The Eq. (7.32) of the DMFC is the following

$$CH_3OH + \frac{3}{2}O_2 \longrightarrow 2H_2O + CO_2$$

1. Eq. (7.7) can be rewritten as follows for the change in Gibbs free energy.

$$\Delta G = \sum (\Delta G_f)_{products} - \sum (\Delta G_f)_{reactants}$$

The chemical equation of the methanol FC is used to modify the above equation.

$$\Delta G = \left[(\Delta G_f)_{CO_2} + 2(\Delta G_f)_{H_2O}\right] - \left[(\Delta G_f)_{CH_3OH} + \frac{3}{2}(\Delta G_f)_{O_2}\right]$$

By taking the values of the Gibbs free energy of the reactants and the products from the given table

$$\Delta G = [-394.36 + 2(-237.13)] - [-166.27 + 0]$$

$$\Delta G = -702.35\text{kJ/mol}$$

2. Eq. (7.7) can be rewritten as the following for the change in enthalpy of the reaction

$$\Delta H = \sum (h_f)_{products} - \sum (h_f)_{reactants}$$

The above equation can be written for Eq. (7.18) as follows

$$\Delta H = \left[(h_f)_{CO_2} + 2(h_f)_{H_2O}\right] - \left[(h_f)_{CH_3OH} + \frac{3}{2}(h_f)_{O_2}\right]$$

By taking the values of the change in enthalpy of the reactants and the products from the given table

$$\Delta H = [-393.5 + 2(-286.02)] - [-239 + 0]$$

$$\Delta H = -726 \text{ kJ/mol}$$

3. The conversion efficiency of the Gibbs free energy into useful work is determined by Eq. (7.14)

$$\eta = \frac{\Delta G}{\Delta H} = \frac{-702.35 \text{ kJ/mol}}{-726 \text{ kJ/mol}}$$

$$\eta = 96.7\%$$

7.7.2 Phosphoric acid fuel cell

PAFCs are similar in operation to the PEMFC and the DMFC except the electrolyte. Phosphoric acid is used as an electrolyte in PAFCs. From the hydrogen storage tank, hydrogen gas enters the FC from the anode side. Porous and thin gas diffusion layers at the anode and cathode provide a suitable flow for the reactant gases into the catalyst layer. It also provides corrosion resistance and thermal and electrical conductivity. A catalyst of platinum mixed with an ion-conducting polymer ionomer is sandwiched between the anode gas diffusion layer and the membrane that helps in splitting the incoming hydrogen molecules into electrons and protons. The protons diffuse over the electrolyte to the cathode gas diffusion layer, whereas the electrons move toward the cathode through the anode and the external DC load. Platinum on the cathode catalyst layer helps the reduction of oxygen molecules into two oxygen atoms where they attract two protons coming through the electrolyte and two electrons coming from the cathode to produce water. The structural diagram of the PAFC is shown in Fig. 7.4.

The oxidation reaction of hydrogen taking place at the anode is shown in Eq. (7.35) where the hydrogen molecules are split into two protons and two electrons. Eq. (7.36) shows the reduction at the cathode combining the oxygen, proton, and electrons to make water.

$$\text{Anode: } 2H_2 \longrightarrow 4H^+ + 4e^- \tag{7.35}$$

$$\text{Cathode: } O_2 + 4H^+ + 4e^- \longrightarrow 2H_2O \tag{7.36}$$

At the operating temperature range of 160°C–220°C, the by-product water can be used for a combined heat and power process that raises the efficiency of the PAFCs from 40% to 70%. The other advantage of the PAFCs is the flexibility of the fuel as CO_2 and 1.5% CO can also be used as a fuel.

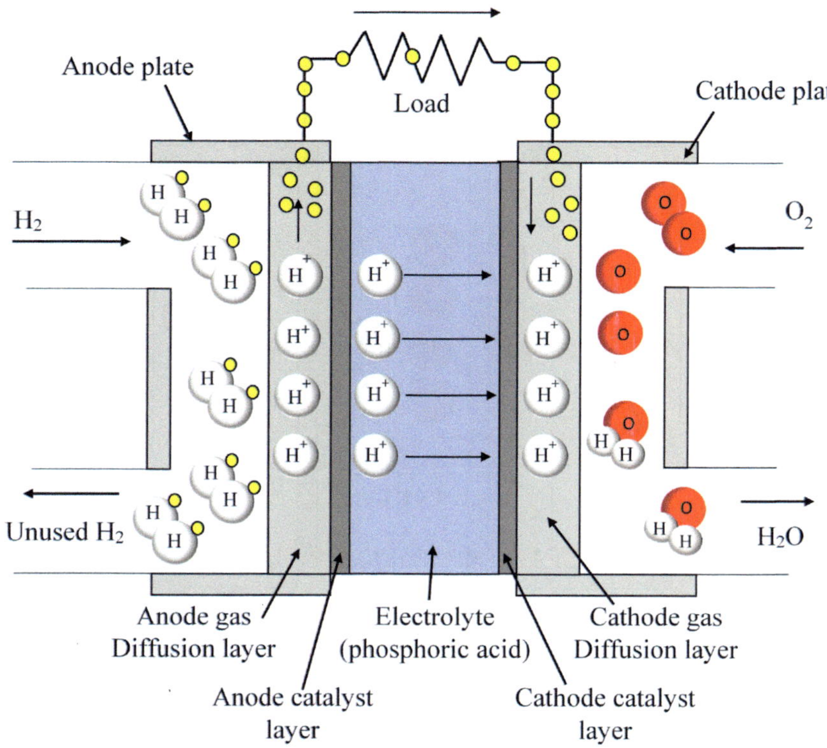

FIGURE 7.4 Phosphoric acid fuel cell.

7.7.3 Alkaline fuel cell

Alkaline FCs (AFCs) are one of the most developed and widely used technologies. They are mostly used in space shuttles for electrifying the onboard system and for the potable water needs. Their name is derived from the definition of alkali that is the base soluble in water. As the electrolyte used in this FC is water-based potassium hydroxide, so it is called an AFC. The working procedure of the AFC is shown in Fig. 7.5. The AFC consists of two electrodes working as anode and cathode, anode and cathode gas diffusion layers, and anode and cathode catalyst layers of platinum. Instead of the proton exchange membrane, it has an aqueous electrolyte of potassium hydroxide (KOH). Oxygen enters the FC from the cathode side inlet where it collects the electrons coming from the anode to the cathode through the external load and the water from the electrolyte. The end product of this reaction is the hydroxide ion (OH^-). Hydrogen enters the FC from the anode inlet where it reacts with the OH^- ions coming from the cathode through the KOH electrolyte to form water and electrons. These electrons generate DC electricity by moving to the external load and the cathode. The equation of the oxidation reaction of

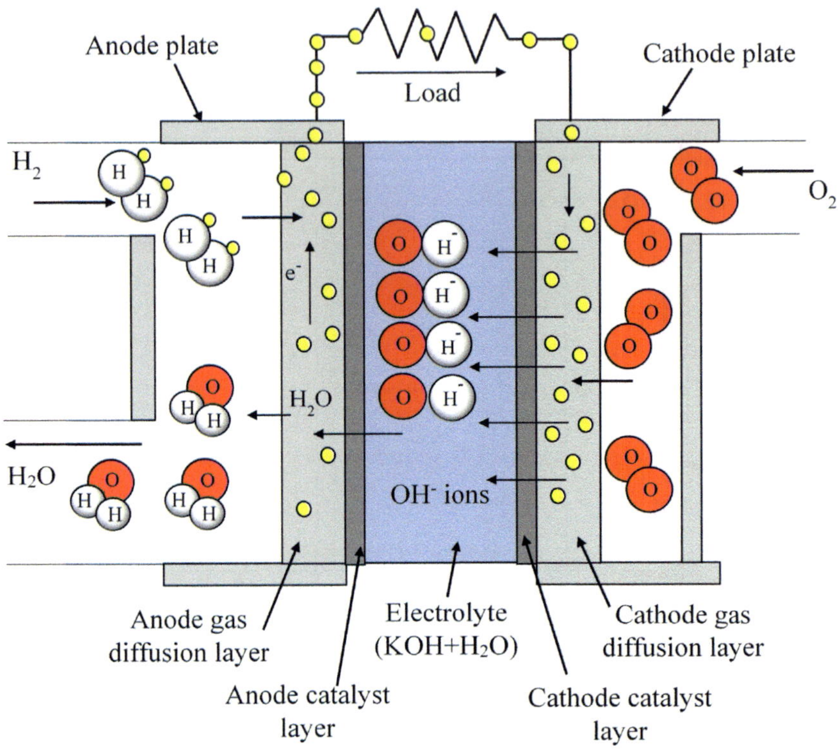

FIGURE 7.5 Alkaline fuel cell.

hydrogen at the anode and the equation of the reduction reaction of oxygen at the cathode are shown in Eqs. (7.37) and (7.38), respectively.

$$\text{Anode: } 2H_2 + 4OH^- \longrightarrow 4H_2O + 4e^- \tag{7.37}$$

$$\text{Cathode: } O_2 + 4e^- + 2H_2O \longrightarrow 4OH^- \tag{7.38}$$

The advantage of the AFCs is the choice to use nonprecious metals like nickel and nickel oxide as a catalyst. The other advantage is the ability to start quickly even at low temperatures. However, the application of the AFCs is limited to the small power applications where pure hydrogen and oxygen are used. If impure fuel containing CO_2 is used, it reacts with the KOH and forms precipitates of the carbonates at the electrodes that further hinder the oxidation and reduction reactions.

7.7.4 Molten carbonate fuel cell

Molten carbonate FCs (MCFCs) are FCs working at a high temperature of the order of 600°C. Like other FCs, MCFCs consist of two electrodes named anode and cathode, anode and cathode gas diffusion layers, anode and cathode catalyst layers of platinum, and molten lithium and potassium carbonate electrolyte. MCFCs have an internal reforming facility so they do not need any external reformer to avoid poisoning by CO_2 and CO. CO_2 itself is used as a fuel at the cathode where it is reacted with oxygen and electrons to form the carbonate ion (CO_3^{-2}).

CO_2 and O_2 enter the FC from the cathode where they react with the electrons coming from the anode through the externally connected DC load and make CO_3^{-2}. The carbonate ions travel through the molten carbonate electrolyte and reach the anode. The cathode reaction is shown in Eq. (7.39). On the other side, H_2 enters through the anode inlet where it reacts with the carbonate ion (CO_3^{-2}) coming from the electrolyte to form water, carbon dioxide, and electrons. The water is collected at the second outlet of the FC at the anode side. Electrons move to the external load generating DC electricity and the carbon dioxide is sent back to the cathode where it is reused in the cathode reaction as described earlier. The anode and cathode reactions are shown in Eqs. (7.39) and (7.40) respectively. The overall working of a MCFC is described in Fig. 7.6.

$$\text{Cathode: } O_2 + 2CO_2 + 4e^- \longrightarrow 2CO_3^{-2} \tag{7.39}$$

$$\text{Anode: } H_2 + CO_3^{-2} \longrightarrow H_2O + CO_2 + 2e^- \tag{7.40}$$

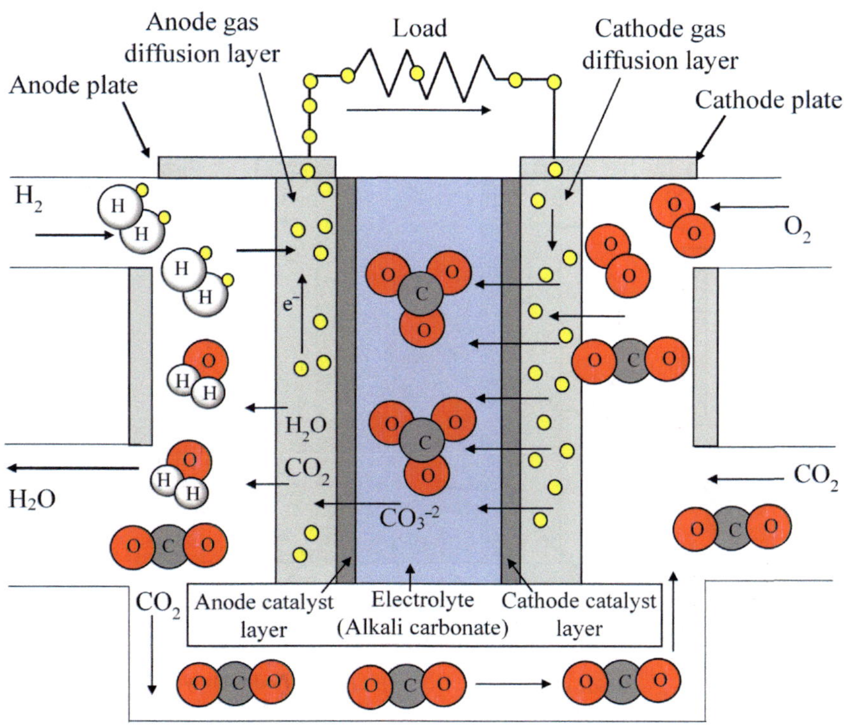

FIGURE 7.6 Molten carbonate fuel cell.

The key advantageous characteristics of the MCFCs are as follows: firstly, the high operating temperature (600°C–700°C) of the MCFCs makes them cost-effective from the aspect of catalyst. There is no need to use precious metals like Pt as a catalyst and nonprecious and inexpensive metals like Ni can be used as a catalyst. Secondly, they are not disposed to the poisoning gases like CO and CO_2. Oxides of carbon or coal-derived gases can be used as a fuel in MCFCs. Thirdly, they can attain high conversion efficiency of the order of 50%–60% that can be increased to 80% when the system is used in the cogeneration process.

7.7.4.1 Solid oxide fuel cell

Solid oxide FCs (SOFCs) are FCs working at a high temperature of the order of 800°C–1000°C. SOFC is considered highly efficient, nearly 80% when used with cogeneration. Like other FCs, SOFC consists of two electrodes working as anode and cathode, anode and cathode gas diffusion layers, anode and cathode catalyst layers of platinum, and a dense ceramic electrolyte separating the anode and cathode. The working principle and the structure of the SOFC are shown in Fig. 7.7.

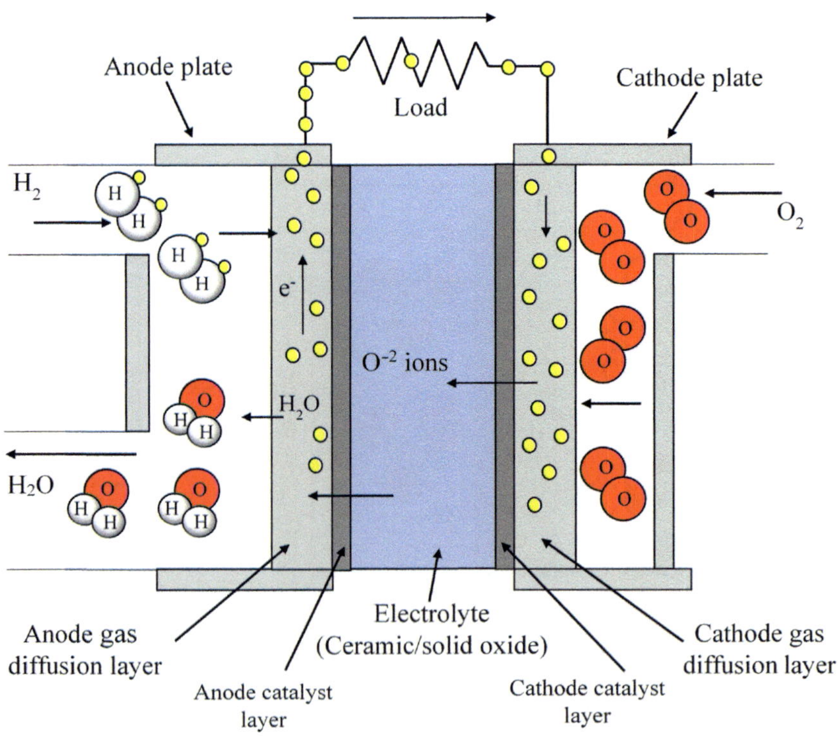

FIGURE 7.7 Solid oxide fuel cell.

From the single inlet at the cathode side, oxygen is passed to the cathode where its reduction occurs by the electrons approaching from the anode through the external DC load to form O^{-2}. These oxygen ions travel through the electrolyte and reach the anode where they oxidize the hydrogen gas coming from the anode inlet. As a result of this reaction, water molecules and electrons are produced. The water is collected out of the FC, whereas the electrons move to the anode to entertain the external DC load and reach the cathode. The anode and cathode reactions are shown in Eqs. (7.41) and (7.42), respectively

$$\text{Cathode: } O_2 + 4e^- \longrightarrow 2O^{-2} \tag{7.41}$$

$$\text{Anode: } H_2 + O^{-2} \longrightarrow H_2O + 2e^- \tag{7.42}$$

The major advantage of the SOFC is the variety of fuels like light hydrocarbons, that is, methanol, ethanol, methane, propane, and butane, that can perform the functioning of the internal reforming facility at the anode. The high operating temperature of the order of 800°C–1000°C makes it useful for the cogeneration industry, which increases the efficiency of the FC. A comprehensive summary of the different types of the FCs is given in Table 7.2

Example 7.2: The FC company is to make a FC of 3 MW operating at 500 mV and a current density of 450 mA/cm^2.

(a) What should be the area of the FC?

(b) If the area of a single cell is 2 m^2 and each stack contains 300 cells, how many stacks are needed to generate 3 MW?

Solution

(a) The area of the FC with given parameters is determined by dividing the current by the current density. The unknown parameter can be determined by using the relation

$$P = V \times I$$
$$I = \frac{P}{V} \Rightarrow \frac{3MW}{500mV} = 6MA$$

Current density is defined as the current per unit area.

$$J = \frac{I}{A}$$

$$A = \frac{I}{J} \Rightarrow \frac{6MA}{450\ \text{mA/cm}^2}$$

$$A = 13333333\ \text{cm}^2$$

(b) First, we calculate the number of cells required by dividing the total area by the area of a single cell.

TABLE 7.2 Summary of the types of fuel cells.

	Parameters						
Fuel Cell	**Electrolyte**	**Fuel**	**Catalyst**	**Operating temperature**	**Efficiency**	**Anode reaction**	**Cathode reaction**
PEMFC	Nafion polymer	H_2	Platinum	60°C–80°C	40%–50%	$H_2 \longrightarrow 2H^+ + 2e^-$	$\frac{1}{2}O_2 + 2H^+ + 2e^- \longrightarrow H_2O$
DMFC	Nafion polymer	Methanol–water	Platinum	110°C	<40%	$CH_3OH + H_2 \longrightarrow 6H^+ + 6e^- + CO_2$	$\frac{3}{2}O_2 + 6H^+ + 6e^- \longrightarrow 3H_2O$
AFC	$KOH + H_2O$	H_2	Non platinum group metals	Up to 230°C	70%	$2H_2 + 4OH^- \longrightarrow 4H_2O + 4e^-$	$O_2 + 4e^- + 2H_2O \longrightarrow 4OH^-$
PAFC	Liquid phosphoric acid	H_2	Platinum	160°C–220°C	40%–80% with cogeneration	$2H_2 \longrightarrow 4H^+ + 4e^-$	$O_2 + 4H^+ + 4e^- \longrightarrow 2H_2O$
MCFC	Alkali carbonates	H_2 and other hydrocarbons	Nonprecious metals	600°C–700°C	60%–80% with cogeneration	$H_2 + CO_3^{-2} \longrightarrow H_2O + CO_2 + 2e^-$	$O_2 + 2CO_2 + 4e^- \longrightarrow 2CO_3^{-2}$
SOFC	Solid ceramic/ solid oxide	H_2 and other hydrocarbons	Nonprecious metals	800°C–1000°C	60%	$H_2 + O^{-2} \longrightarrow H_2O + 2e^-$	$O_2 + 4e^- \longrightarrow 2O^{-2}$

No. of cells $= \frac{13333333\ \text{cm}^2}{2\text{m}^2}$

No. of cells = 666 cells

No. of stacks $= \frac{666\ \text{cells}}{250\ \text{cells/stack}} = 2.66 \simeq 3\text{stacks}$

Problems

Problems 1–10 contain four answer options A, B, C, and D. Choose the correct answer.

1. Enthalpy change of the formation of water is
   A. -286kJ/mol
   B. -285kJ/mol
   C. $-287\frac{\text{kJ}}{\text{mol}}$
   D. -288kJ/mol
2. Entropy of the formation of water is
   A. -0.153kJ/mol
   B. -0.163kJ/mol
   C. -0.173kJ/mol
   D. -0.183kJ/mol
3. Gibbs free energy for the formation of water is
   A. -227kJ/mol
   B. -237kJ/mol
   C. -247kJ/mol
   D. -257kJ/mol
4. The correct relation for the Gibbs free energy is
   A. $\Delta G = \Delta H + T\cdot\Delta S$
   B. $\Delta G = \Delta H - T\cdot\Delta S$
   C. $\Delta G = \Delta H - T\cdot\Delta H$
   D. $\Delta G = \Delta T - T\cdot\Delta S$
5. Terminal voltage of the stack of the FC is given by
   A. $V_T = N(E_T - L)$
   B. $V_T = N(E_T + L)$
   C. $V_T = (E_T - L)$
   D. $V_T = N(E_T - \eta_{ohmic})$
6. The efficiency of a DMFC is
   A. 95.7 %
   B. 96.7 %
   C. 97.7 %
   D. 98.7 %
7. The anode reaction of DMFC is
   A. $CH_3OH + H_2 \longrightarrow 6H^+ + 6e^- + CO_2$
   B. $H_2 + CO_3^{-2} \longrightarrow H_2O + CO_2 + 2e^-$
   C. $\frac{3}{2}O_2 + 6H^+ + 6e^- \longrightarrow 3H_2O$
   D. None of the above

8. Enthalpy of formation of methanol is
   A. −726kJ/mol
   B. +726kJ/mol
   C. −736kJ/mol
   D. +736kJ/mol
9. Which is the incorrect information about the FC?
   A. The operation of the FC is quite noisy
   B. FCs are modular
   C. The efficiency of the FC is high
   D. The emissions from the FCs are quite low.
10. Which of the following is not an example of the FC?
    A. Hydrogen−oxygen FC
    B. Hexanone−oxygen FC
    C. DMFC
    D. PAFC
11. What is common between a solar photovoltaic cell and a FC?
12. What type of losses are involved in a FC?
13. Calculate the Gibbs free energy of the following reaction.
    $$CH_3OH + \frac{3}{2}O_2 \longrightarrow 2H_2O + CO_2$$
14. What is proton exchange membrane and how does it play its role in the working of a FC?
15. What are the most common ways of producing hydrogen to be used in the FC?

Chapter 8

Bioenergy

Muhammad Kamran
Department of Electrical Engineering and Technology, Riphah International University, Pakistan

8.1 Introduction

Since the history of cooking and space heating, biomass has been used in various forms with different technologies. Today biomass has become an environment-friendly source of energy and is expected to be a sustainable economy. To maintain the sustainability of bioenergy, the farmers need to minimize the demand and supply gap at low prices. To understand the dynamics of bioenergy, one must know the bioenergy market along with other sources of renewable energy, different forms of bioenergy, conversion technologies, and the by-products of these conversion technologies. The liquid, solid, and gas fuels that are derived from biomass through the different processes are called biofuels. Gas and liquid biofuels produced through pyrolysis, fermentation, gasification, and transesterification are used for industrial processing and transportation.

The primary source of biomass energy is solar energy. The plants take the carbon dioxide from the atmosphere and water from the soil to form glucose and oxygen in the presence of sunlight. This process of preparing glucose in the presence of sunlight is called photosynthesis. In the dark, the photosynthesis process does not occur. Eq. (8.1) shows the photosynthesis process.

$$6CO_2 + 6H_2O \rightarrow 2C_6H_{12}O_6 + 6O_2 \tag{8.1}$$

Glucose ($C_6H_{12}O_6$) is a carbohydrate that can be converted to cellulose and starch in the plant or even can be converted to proteins and other compounds by combining with nitrogen. The other product oxygen is released into the atmosphere, which is used by the plants and animals for respiration.

8.2 Biomass

Conventional fossil fuels (oil, gas, and coal) are being used to meet global energy demands. With the passage of time and their excessive utilization,

Renewable Energy Conversion Systems. DOI: https://doi.org/10.1016/B978-0-12-823538-6.00002-6

these conventional energy sources are being depleted rapidly and they will not be replenished in the near future. The excessive use of conventional energy sources has put a serious alarming impact not only on the environment but also on human health. The alternative sources of energy not only replenish themselves but also have the minimum health hazards. Agricultural biomass that is considered a renewable energy source is abundantly available on Earth.

Four types of biomass materials—vegetable oil, sugar and starch, lignocellulosic materials, and wet biomass—can be converted into four types of energy sources—biofuels, chemicals, electricity, and heat—by using different corresponding technologies. Vegetable oils and animal fats are converted into biodiesel through transesterification. The sugar and starch-containing biomass are converted into ethanol through hydrolysis and fermentation. Ethanol is used as a chemical and a biofuel. Lignocellulosic biomass can be converted into (1) ethanol through the hydrolysis–fermentation process, (2) hydrocarbons and biofuels through pyrolysis, (3) producer gas through gasification, and (4) biomass pallets through the palletization process. The producer gas and the pallets can be used to generate electricity or for heating purposes. The wet biomass is converted into biogas through anaerobic digestion [1]. The types of biomass, available conversion technologies, and the end products are detailed in Fig. 8.1.

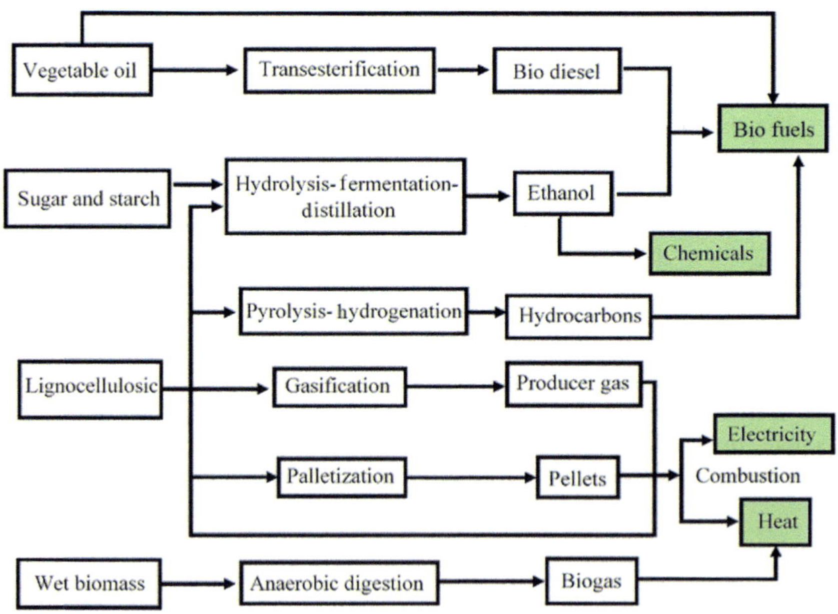

FIGURE 8.1 Types of biomass, conversion technologies, and end products.

The sources of these biomass resources are sugar cane top, sugar cane trash, bagasse, wheat straw, cotton shells, cotton stalks, rice husk, animal dung, animal manure, and municipal solid waste. These biomass resources can be grouped as follows.

- Agricultural waste, crop residues and crop processing residues, energy crops, and animal wastes and manures.
- Industrial waste and municipal solid waste.
- Residues and waste from forests.

These biomass materials are characterized by the proximate analysis in which moisture content, ash, volatile matter, and fixed carbon are measured. In the ultimate analysis of the biomass the elemental composition of the solid biomass is determined. The major elements that are determined in the ultimate analysis are carbon, hydrogen, nitrogen, sulfur, and oxygen [2].

- *Moisture content*: moisture content is the percentage moisture in a sample if that sample of biomass is heated just slightly above the boiling point of water (105°C). The heat at this temperature evaporates the water from the sample of the biomass, thus changing the weight of the biomass. When it stops changing the difference between the initial weight and the final weight expresses the moisture content as the weight loss. It is expressed as the percentage of the biomass material.
- *Volatile matter*: volatile matter is that which is easily burnt out in the presence of oxygen and is determined as mass loss.
- *Ash contents:* when the biomass is incinerated, the remaining material with no energy value is known as the ash of the biomass. It is made up of inorganic matters. If the value of the ash content in the specific biomass is beyond the acceptable limits, it causes problems in the thermochemical processes like pyrolysis, combustion, and gasification.
- *Fixed carbon:* a combustible remaining solid residue when a sample of the biomass is burnt at a temperature of 750°C for 7 minutes. When the water, ash contents, and the volatile matter are removed from the biomass, the remaining is the fixed carbon, as shown in Eq. (8.2).

$$\text{Fixed carbon} = 100 - (\%H_2O + \%\text{Ash} + \%\text{VM}) \tag{8.2}$$

Example 8.1

Perform the proximate analysis on the sample of 1 g of biomass.

Solution

To perform the proximate analysis on the given sample of the biomass, we need to find the moisture content, ash content, volatile matter, and the fixed carbon.

1. *Moisture contents:* we start by heating the 1 g biomass sample at a temperature of 105°C until the mass stops changing. The moisture in the

sample will evaporate and we measure the initial weight and the final weight of the sample. The initial weight was 1 g while the final weight is 0.90 g, hence the moisture content in the biomass sample is:

$$\text{Moisture contents} = 1 - 0.90\ \text{g}$$

$$\text{Moisture contents} = 0.10\ \text{g or } 10\ \%\ \text{moisture weight}$$

2. *Ash content:* the sample remaining from the first test of moisture content is burnt in the air until its weight stops changing. The final weight will be the weight of the ash that has no heating value and is of no importance. The final weight measured was 0.22 g.

$$0.22\ \text{g of ash} = 22\ \text{wt. percent ash}$$

3. *Volatile matter:* the volatile matter is that which is easily burnt out in the absence of oxygen. Another sample of the biomass is burnt for 7 minutes at a temperature of 750°C. The initial and the final mass of the biomass are recorded. From the initial 1 g biomass at the end 0.60 g was left.

$$\text{Volatile matter} = 1 - 0.60\ \text{g}$$

$$\text{Volatile matter} = 0.40\ \text{g or } 40\ \%\ \text{volatile matter}$$

4. *Fixed carbon:* fixed carbon of the biomass is determined by the following relation

$$\text{Fixed carbon} = 100 - (\%H_2\text{O} + \%\text{Ash} + \%\text{VM})$$

$$\text{Fixed carbon} = 100 - (10\% + 22\% + 40\%)$$

$$\text{Fixed carbon} = 28\%$$

Energy density in biomass is defined as the amount of energy contained. Table 8.1 shows the storage density and the energy density of different

TABLE 8.1 Density and energy density of different biomass types.

Biomass type	Energy density (GJ/m^3)	Density (kg/m^3)
Bio-oil	20	1200
Torrefied pallets	15	800
Pellets	9.8–14.0	550–700
Briquettes	6.4	350
Ground biomass	3.6	200
Baled biomass	2.8–4.7	160–255

biomass resources. Energy density of a specific biomass is increased by making pellets and briquettes.

8.2.1 Palletization

A major problem with biomass is its low density that makes it difficult to handle, transport, and store. To tackle these problems, pallets and briquettes are made. These are made from crop stalks, sawdust, crops residues, rice husk, cotton shells, bagasse, peanut shells, and other agricultural wastes. Biomass palletization is a process of storing the high density and solid energy carrier from biomass. The palletization process requires some pretreatment and posttreatment processes. In the pretreatment process, biomass is kept in a safe area to avoid moisture and impurities. Metal chips and other impurities are removed from the biomass to avoid damage to the equipment. The bigger sized biomass is reduced to the acceptable size. The size of the biomass must be less than the hole of the dye of the pallet mills. In the next step, uniformly sized biomass material is dried as wet biomass/pallets would cause less efficient burning and the production of the smoke and hydrocarbons. The major advantage of the pallets and briquettes are the following:

- Briquettes increase the combustion efficiency of the fuel
- A decrease in volume of the waste biomass.
- Easy and continuous supply of fuel to the furnace
- Environmental friendly
- Less danger of fire risk in facility and storage.

8.3 Biogas

Agricultural waste, animal dung, animal manure, and sometimes human excreta can be converted into biogas and biofertilizer by the process of anaerobic digestion. In this process, biodegradable complex materials (fats, proteins, and carbohydrates) are decomposed by the acidic and methane bacteria into simple molecules (fatty acids, amino acids, and sugars) that are soluble in water. This process is termed methane fermentation and is carried out in the following four steps.

8.3.1 Anaerobic digestion process

8.3.1.1 Hydrolysis

Hydrolysis is the first step of anaerobic digestion. In hydrolysis, long-chain complexed polymers of organic matter that are mostly not usable in their current form are decomposed into simple lower sugars, fatty acids, and amino acids. The rate of the overall process is hindered by the hydrolysis process, especially when the waste material is solid. The reaction associated with this process is shown in Eq. (8.3), in which the cellulosic material is hydrolyzed by the addition of water to produce glucose [3].

$$C_6H_{10}O_5 + H_2O \rightarrow C_6H_{12}O_6 + H_2 \tag{8.3}$$

8.3.1.2 Acidogenesis

In the acidogenesis process, the biodegradable materials like fats, protein, and bacteria are further decomposed by the fermentation bacteria. As a result of the acidogenesis process, volatile fatty acids, carbon dioxide, hydrogen sulfide, and ammonia are produced. The chemical equations associated with the acidogenic process are Eqs. (8.4)–(8.6).

$$C_6H_{12}O_6 \rightarrow 2C_2H_5\,OH + 2CO_2 \tag{8.4}$$

$$C_6H_{12}O_6 + 2H_2 \leftrightarrow C_2H_5\,COOH + 2H_2O \tag{8.5}$$

$$C_6H_{12}O_6 \leftrightarrow 3CH_3COOH \tag{8.6}$$

8.3.1.3 Acetogenesis

The product of the acidogenesis is further decomposed by the acetogens to produce hydrogen, carbon dioxide, and acetic acid. Generally, the production of acetate is called acetogenesis. The equations associated with this process are Eqs. (8.7)–(8.9).

$$C_2H_5\,COO^- + 3H_2O \leftrightarrow CH_3COO^- + H^+HCO_3^- + 3H_2 \tag{8.7}$$

$$C_6H_{12}O_6 + 2H_2O \leftrightarrow 2CH_3COOH + 2CO_2 + 4H_2 \tag{8.8}$$

$$C_2H_5OH + 2H_2O \leftrightarrow CH_3COO^- + 3H_2 + H^+ \tag{8.9}$$

8.3.1.4 Methanogenesis

In the fourth and the last step of the anaerobic digestion, the methanogens decompose the hydrogen, carbon dioxide, and acetic acid into methane and carbon dioxide. The equations associated with this process are Eqs. (8.10)–(8.12). The four steps of anaerobic digestion are shown in Fig. 8.2.

$$3CH_3COOH \rightarrow CH_4 + CO_2 \tag{8.10}$$

$$CO_2 + 4H_2 \rightarrow CH_4 + 2H_2O \tag{8.11}$$

$$2C_2H_5OH + CO_2 \rightarrow CH_4 + 2CH_3COOH \tag{8.12}$$

The percentage composition of the output products as a result of the anaerobic digestion of the biomass material is as follow: CH_4 is 50%–75%, CO_2 is 25%–50%, N_2 is 0%–10%, H_2 is 0%–1%, H_2S is 0%–3%, and O_2 is 0%–2%. It can be seen that CH_4 is the major constituent of the biogas.

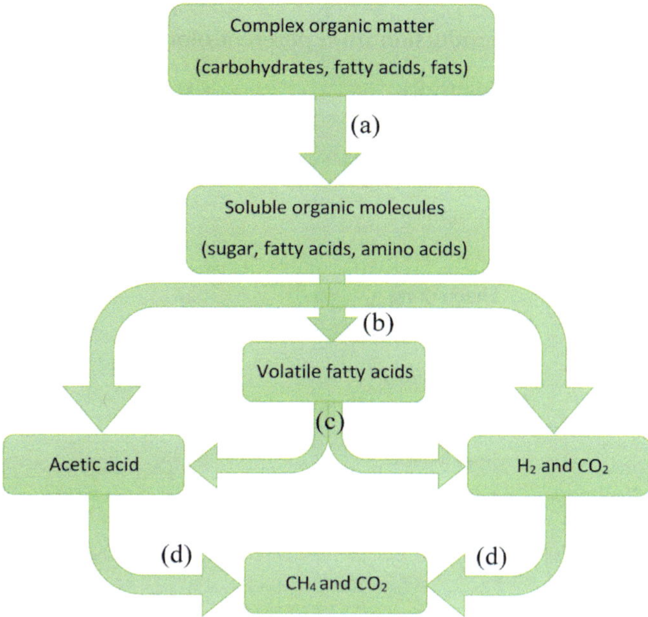

FIGURE 8.2 Process of anaerobic digestion.

The percentage of the constituents of the biogas may vary depending on the feedstock composition used in the digestive process. The percentage of methane in various feedstocks is given in Table 8.2. The feedstock that could be used as a substrate in the biogas digester is as follows:

- Food processing wastes
- Waste feed
- Slaughterhouse waste
- Corn silage
- Energy crops
- Industrial wastes
- Slurries
- Crop residues
- Livestock manure
- By-products from ethanol and biodiesel production

The process of biodigestion is carried out in the digester, as shown in Fig. 8.3. The feedstock is fed to the digester through the dung feeder connected to the digester through a cement pipe. The anaerobic digestion process starts in the digester and the produced methane gas is collected in the gas collector dome. The gas is collected and is used for heating and cooking purposes. The animal dung and other material after the production of biogas are collected from the digester and used as fertilizer in fields [4].

TABLE 8.2 Methane production from different biomass feedstocks.

Sr. no.	Feedstock	CH_4 composition (%)
1	Cattle manure	50–60
2	Poultry waste	68
3	Corn silage	54.5
4	Barley straw	77
5	Sheep dung	65
6	Grass	84
7	Wheat straw	78.5
8	Horse dung	66
9	Beet leaves	84.8
10	Dried leaves	58
11	Pig manure	60

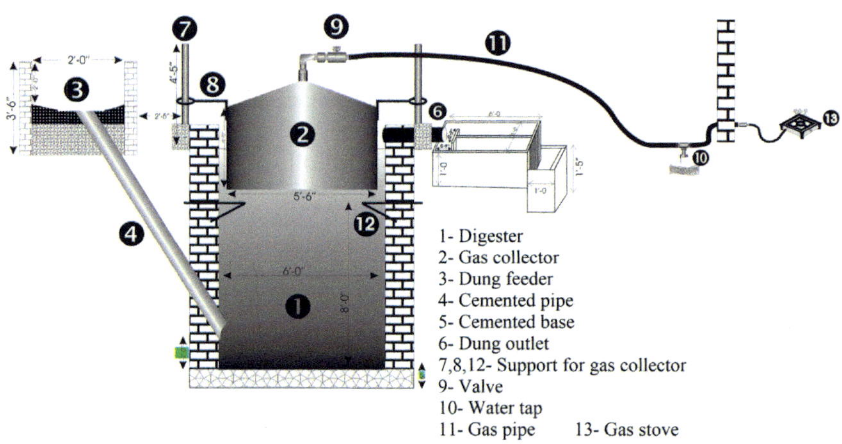

FIGURE 8.3 Biogas plant.

The primary constituents of the biogas are methane and carbon dioxide with a minute amount of nitrogen and hydrogen. The detailed composition of the biogas is methane is 50%–75%, carbon dioxide is 25%–50%, 1%–10% water content, and traces of ammonia, oxygen, hydrogen sulfide, and nitrogen. The amount of methane determines the energy content of the biogas. The higher the concentration of the biogas, the higher will be the energy content.

8.4 Biodiesel

The oil extracted from the plants, animal fats, and recycled grease contains various impurities like free fatty acids, water, odorants, sterols, and phospholipids because of which these plant oils cannot be directly used as fuel. To make it usable as a fuel it requires some modification like esterification. The production of biodiesel consists of the transesterification of alcohol like methanol with oil and fat in the presence of a catalyst. The reaction results in the production of mono-alkyl esters termed biodiesel and glycerin as a by-product. The process of biodiesel production is shown in Fig. 8.4. The oil and fats could be animal fats, vegetable oil, or any other triglyceride material. A strong base like sodium hydroxide (NaOH) or potassium hydroxide (KOH) can be used as a catalyst. The overall reaction of the transesterification of the methanol and triglyceride is shown in Fig. 8.5. The produced biodiesel is similar to conventional diesel in its characteristics [5].

8.4.1 Physical characteristics of biodiesel

8.4.1.1 Flashpoint

The flashpoint of the biodiesel is not the operation-related property but it governs the storage and handling of the fuel. It is the temperature at which the surface of the liquid fuel generates enough vapors to ignite. The

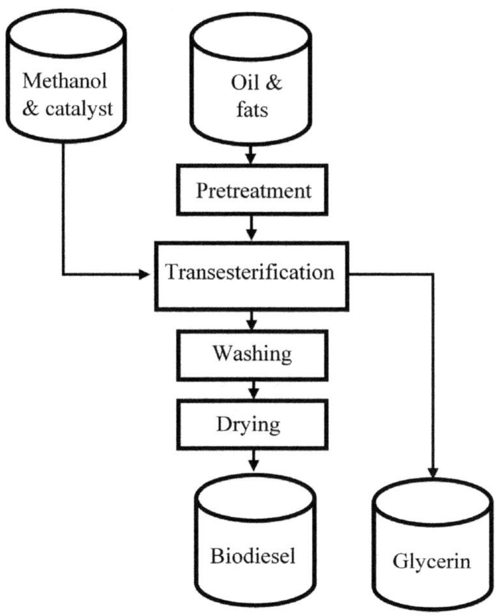

FIGURE 8.4 Transesterification of methanol to form biodiesel.

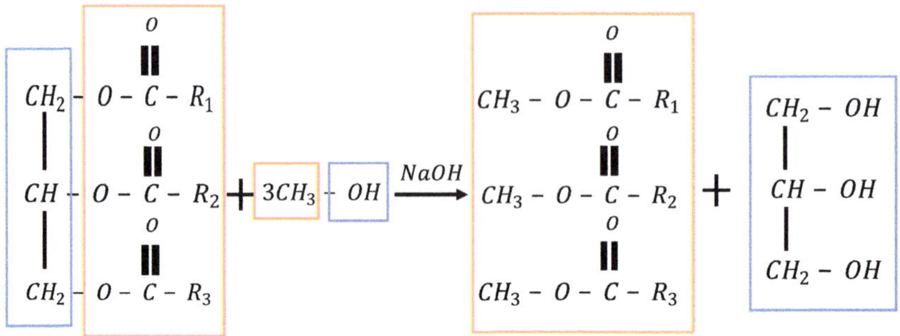

FIGURE 8.5 Chemical bonding equation of biodiesel formation.

flashpoint of the biodiesel is higher (average 150°C) than the fossil diesel (average range 55°C–66°C). The flashpoint of the biodiesel is measured by the close cup method in which the sample is heated at a constant rate in a closed cup and the temperature is regularly measured at the interval of 1°C or 2°C. At each interval, the cup is opened and brought close to the ignition source. If the vapors over the surface of the biodiesel sample get ignited, this temperature will be termed as the flashpoint of the fuel. The fuels with a low flashpoint are carefully stored and handled as they get ignited at a lower temperature [6].

8.4.1.2 Boiling point

The boiling point of biodiesel is the temperature at which the vapor pressure of the biodiesel becomes equal to the pressure of the atmosphere around the biodiesel surface. At this temperature, biodiesel starts converting into vapors. If the atmospheric pressure changes, the boiling point of the fuel also changes.

8.4.1.3 Cloud point

Cloud point is the cold weather characteristic of the biodiesel and is defined as the minimum temperature at which the crystals of the biodiesel start to form. The temperature of the test tube sample of the biodiesel is reduced in steps of 3°C. The temperature is noted at which the crystals or wax or clouds start to form and this temperature is called the cloud point.

8.4.1.4 Pour point

Once the cloud point is achieved, the temperature of the biodiesel sample is further decreased in a step of 1°C until the biodiesel stops moving or it loses its flowing characteristics. Both the cloud point and the pour point are the cold temperature characteristics of the fuel and become significant in the cold regions.

8.4.1.5 Calorific value

Calorific value is defined as the amount of energy generated by burning 1 g of fuel in the presence of oxygen measured as kJ/kg and is calculated by the following relation.

$$\text{Calorific } value = \frac{heat\ produced}{amount\ of\ fuel}$$

The heat produced can be calculated by thermodynamics methodology or can be found experimentally.

$$Q = nC_P\Delta T$$

where Q is heat produced, n is no. of moles of fuel, C_P is heat capacity at constant pressure (1 bar), and ΔT is initial and final temperature difference

Example 8.2

In a boiler, 5.0 kg biodiesel was completely burnt that released 150,000 kJ heat. Calculate the calorific value of the fuel.

Solution

As the calorific value is the amount of heat produced per unit mass of the fuel. So we can use the following formula to determine the calorific value.

$$\text{Calorific } value = \frac{heat\ produced}{amount\ of\ fuel}$$

$$\text{Calorific } value = \frac{150,000\ kJ}{5\ kg}$$

$$\text{Calorific } value = 30,000\ kJ/kg$$

8.5 Hydrogen production

Among renewable energy sources, biomass is the key source of energy in different secondary forms like biodiesel, ethanol, biogas, hydrogen, and heat from the incineration of the biomass residues. Different forms of biomass energy sources are agricultural residues, crop residues, industrial and household biomass waste, and municipal wastes.

Hydrogen is not a source of energy, rather it is a carrier of energy that is not easily available naturally. Energy from another energy source is used to produce hydrogen. If the hydrogen is produced from renewable energy sources its production becomes dependent on the availability of the

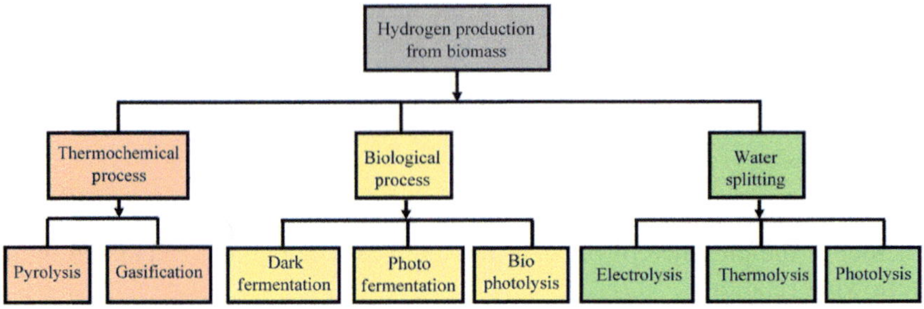

FIGURE 8.6 Hydrogen production from biomass through different processes.

inexhaustible, intermittent, and replenishing renewable energy sources. Because of the following characteristics, hydrogen is considered the ideal energy carrier.

- Hydrogen can be produced by using electricity and electricity can be used to produce hydrogen efficiently.
- Hydrogen is produced by water which is abundantly and freely available.
- Hydrogen can be stored in liquid, gaseous, and metal hydride form.
- Hydrogen is a renewable fuel.
- It can be transported long distances via tankers or dedicated pipelines.
- It can be changed to other forms of energy, for example, in a fuel cell it is converted into electricity.
- It is environmentally friendly as its production, transportation, storage, and conversion into other forms of energy do not involve any emissions.

Biomass is used to generate hydrogen in two ways: (1) thermochemical process and (2) biological process. Fig. 8.6 shows the different thermochemical and biological processes that are used to produce hydrogen from biomass.

8.5.1 Biological processes

8.5.1.1 Biophotolysis

Biophotolysis is the dissociation of water molecules to form hydrogen and oxygen in biological organisms under solar radiation. Biological organisms like cyanobacteria and microalgae are known as autotrophic organisms with the capability of performing photosynthesis. In photosynthesis, the autotrophic organisms absorb the carbon dioxide from the atmosphere and the water from the soil in the presence of solar radiation to generate carbohydrate biomass. This process is called indirect biophotolysis that takes place in two steps. In the second step, the dark fermentation of these carbohydrates is done to produce hydrogen. The reaction of indirect biophotolysis can be described by Eq. (8.13)–(8.15).

$$6H_2O + 6CO_2 + light \rightarrow C_6H_{12}O_6 + 6O_2 \tag{8.13}$$

$$C_6H_{12}O_6 + 2H_2O \rightarrow 4H_2 + 2CH_3COOH + 2CO_2 \tag{8.14}$$

$$2CH_3COOH + 4H_2O + light \rightarrow 8H_2 + 4CO_2 \tag{8.15}$$

The overall reaction can be written as Eq. (8.16)

$$12H_2O + light \rightarrow 12H_2 + 6O_2 \tag{8.16}$$

In direct biophotolysis, water is directly split into hydrogen and oxygen in the presence of solar radiation without any intermediate stage, as shown in Eq. (8.17).

$$2H_2O + light \rightarrow 2H_2 + O_2 \tag{8.17}$$

8.5.1.2 Dark fermentation

Dark fermentation is a type of biological production of hydrogen. Dark fermentation is carried out by obligate anaerobes and facultative anaerobes in the absence of light and oxygen. In dark fermentation, bacteria act on the substrate and generate hydrogen. The substrate for the dark fermentation is lignocellulosic biomass, carbohydrate materials like wastewater from industry, sugar-containing crop residues, and municipal solid waste. In the first step, the pretreatment of the biomass greatly affects the efficiency of the dark fermentation. The other efficiency affecting parameters are the microorganism involved and the sugar content of the substrate. Fig. 8.7 shows the

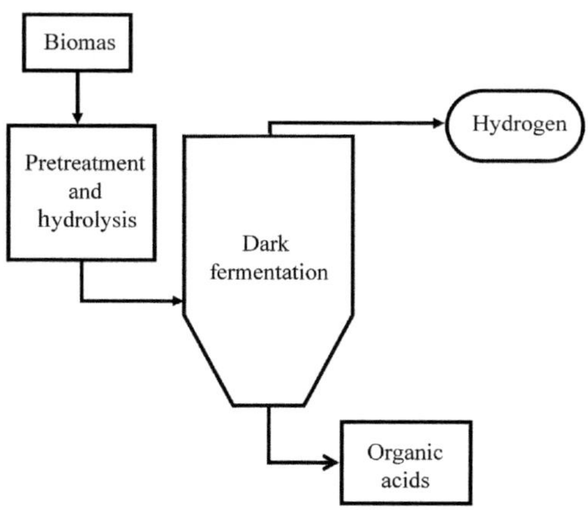

FIGURE 8.7 Dark fermentation of biomass to produce hydrogen.

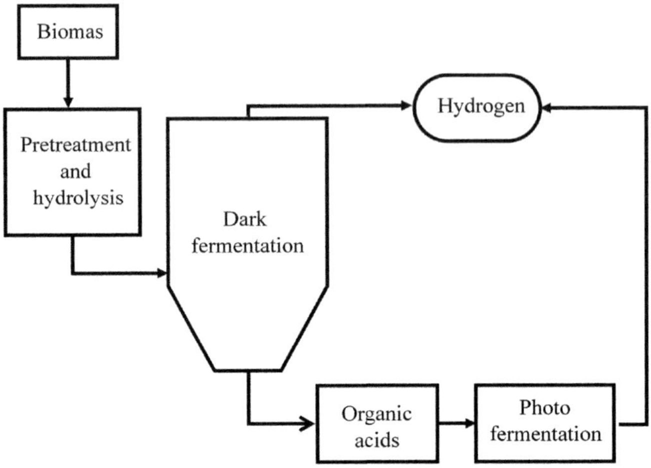

FIGURE 8.8 Photofermentation of biomass to produce hydrogen.

process flow of the dark fermentation. Eq. (8.18) shows the dark fermentation process.

$$C_6H_{12}O_6 + 2H_2O \rightarrow 2CH_3COOH + 2CO_2 + 4H_2 \tag{8.18}$$

8.5.1.3 *Photofermentation*

The conversion of biomass into hydrogen in the presence of sunlight by the photosynthesizing bacteria is known as photofermentation. The performance and the efficiency of photofermentation depend upon the substrate and the photosynthesizing bacteria. The efficiency of the photofermentation is lower than that of dark fermentation because the growth rate of the dark fermentation bacteria is faster than the photosynthesizing bacteria. For the production of the same volume of hydrogen, a large photofermentation reactor would be required. Eq. (8.19) shows the process of photofermentation.

$$CH_3COOH + 2H_2O \xrightarrow{sun\ light} 4H_2 + 2CO_2 \tag{8.19}$$

From the equation of dark fermentation, we can use the by-products as an input for the photofermentation process. Fig. 8.8 shows the integration of dark and photofermentation.

8.5.2 Thermochemical process

8.5.2.1 *Pyrolysis*

In the absence of air and oxygen, biomass is heated at a temperature ranging from 625K to 800K, as a result of which biomass is decomposed into

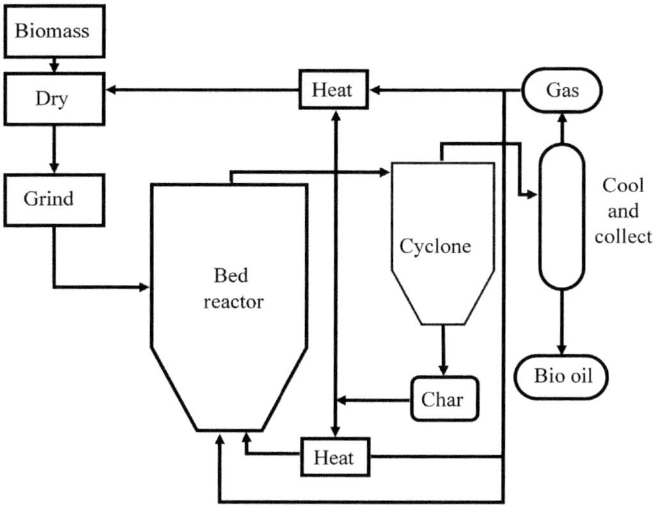

FIGURE 8.9 Pyrolysis of biomass to produce hydrogen.

$$\text{Heat} + Biomass \xrightarrow{pyrolysis} CH_4 + CO + H_2 + other\ products \tag{8.20}$$

$$\text{CH}_4 + H_2O \rightarrow 3H_2 + CO \tag{8.21}$$

$$\text{CO} + H_2O \rightarrow \text{CO}_2 + H_2 \tag{8.22}$$

charcoal, liquid fuel, and gas, as shown in Fig. 8.9. Before sending the biomass in the fluidized bed reactor, it is first dried using the heat generated by the burning of the gas of the pyrolysis process. This dried biomass is pulverized and sent to the bed reactor. The bed reactor could be a fixed bed reactor, fluidized bed reactor, or a bubbling fluidized bed reactor [7].

The dried, and ground biomass enters from one end and the heat enters from the other end. The bubbles of the hot gases move upward where they come in contact with the solid biomass coming downward. The fluidization process of the pulverized biomass by the heat generated from the burning of the gas generates oil vapors, char, and gas. This mixture is sent to the cyclone where the char is collected at the bottom and the remaining is sent for cooling where the gas is collected in the upper chamber and the condensed liquid bio-oil is collected at the bottom chamber. The gas and the char produced as a result of the pyrolysis process are used to provide the heat required to carry out the process.

Eq. (8.20) shows the production of hydrogen gas, methane, and carbon monoxide along with bio-oil and the biochar product as a result of the pyrolysis process of biomass. The methane gas and the carbon monoxide can further be processed to generate hydrogen gas as shown in Eq. (8.21) and (8.22). The

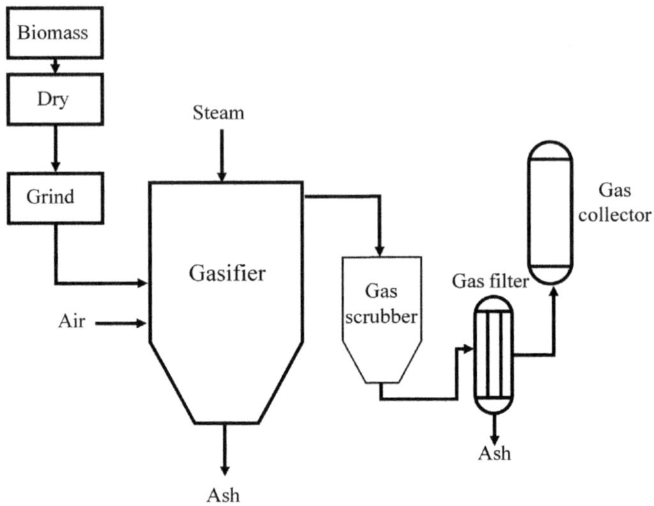

FIGURE 8.10 Gasification of biomass to produce hydrogen.

production of hydrogen apart from the direct reaction can be increased by using different catalysts like $CaCO_3, Na_2CO_3, K_2CO_3, ZnCl_2$. The production of hydrogen can also be increased by raising the temperature of the reaction.

8.5.2.2 Gasification

Biomass gasification is the decomposition of the higher polymers of biomass into lighter molecules by the incomplete combustion of the biomass in the presence of oxidizing agents producing hydrogen, carbon monoxide, and the traces of methane gas. This gasification process takes place at a temperature of 114°C–500°C. Steam and air are used as the gasification agents that oxidize the biomass polymer chains. As a result of the gasification process, carbon monoxide, carbon dioxide, hydrogen gas, methane gas, tar, and char are produced. Eq. (8.23) shows the conversion of biomass into hydrogen gas by the process of gasification. Fig. 8.10 shows the most widely used configuration of biomass gasification. The other products are char and tar.

$$\text{Biomass} + O_2 + H_2O(steam) \xrightarrow{gasification} CH_4 + CO_2 + CO + H_2 + other\ products \tag{8.23}$$

8.5.3 Water-splitting

8.5.3.1 Electrolysis

Among all the techniques of water-splitting to generate hydrogen, electrolysis is the simplest and the most efficient one, adopted more than 200 years ago. Water electrolysis is the process of splitting water into oxygen and hydrogen by passing electricity through the electrolytic cell. The cell consists of an anode, cathode, and electrolyte. The produced hydrogen is used as a fuel in fuel cells. The minimum potential required for the electrolysis of water is 1.23 V. The reaction associated with the electrolysis of water is given by Eq. (8.24).

$$\mathrm{H}_2O + electrical\ energy \Rightarrow \mathrm{H}_2 + \frac{1}{2}O_2 \tag{8.24}$$

In an electrolytic cell, the potential is applied across the anode and cathode dipped in water and electrolyte solution. On the occurrence of the reaction, hydrogen is produced at the cathode, and oxygen is produced at the anode. The produced gas moves against the atmosphere and the system does work in doing so that is denoted as positive. We calculate the enthalpy of the reaction given in Eq. (8.25) as follows: change in enthalpy (ΔH) of a chemical reaction is found by deducting the enthalpy of formation of reactants from the enthalpy of formation of the products using Eq. (8.26).

$$\Delta H = \sum (\mathrm{h_f})_{\mathrm{products}} - \sum (\mathrm{h_f})_{\mathrm{reactants}} \tag{8.25}$$

$$\Delta H = (\mathrm{c} \cdot \mathrm{h_f})_{\mathrm{product\ 1}} + (\mathrm{d} \cdot \mathrm{h_f})_{\mathrm{product\ 2}} - (\mathrm{a} \cdot \mathrm{h_f})_{\mathrm{reactant\ 1}} - (\mathrm{b} \cdot \mathrm{h_f})_{\mathrm{reactant\ 2}} \tag{8.26}$$

where ΔH is change in enthalpy, a is coefficient of reactant 1 from the balanced chemical equation, b is coefficient of reactant 2 from the balanced chemical equation, c is coefficient of product 1 from the balanced chemical equation, d is coefficient of product 2 from the balanced chemical equation, and $\mathrm{h_f}$ is enthalpy of formation.

Change in enthalpy of the formation of water is given in Eq. (8.26) and can be determined using the above-stated procedure. Eq. (8.25) can be redrawn as Eq. (8.27) where water is the product, hydrogen, and oxygen are the reactants.

$$\Delta H = (\mathrm{h_f})_{\mathrm{H_2}} + \frac{1}{2}(\mathrm{h_f})_{\mathrm{O_2}} - (\mathrm{h_f})_{\mathrm{H_2O}} \tag{8.27}$$

$$\Delta H = 0 + 0 - (-286)$$

$$\Delta H = 286\mathrm{kJ/mol}$$

Change in entropy (ΔS) of a chemical reaction is found by deducting the entropy of formation of reactants from the entropy of formation of the products using Eq. (8.28).

$$\Delta S = \sum (\mathrm{S_f})_{\mathrm{products}} - \sum (\mathrm{S_f})_{\mathrm{reactants}} \tag{8.28}$$

$$\Delta S = (\mathrm{c} \cdot \mathrm{S_f})_{\mathrm{product\ 1}} + (\mathrm{d} \cdot \mathrm{S_f})_{\mathrm{product\ 2}} - (\mathrm{a} \cdot \mathrm{S_f})_{\mathrm{reactant\ 1}} - (\mathrm{b} \cdot \mathrm{S_f})_{\mathrm{reactant\ 2}} \tag{8.29}$$

where ΔS is change in entropy, a is coefficient of reactant 1 from the balanced chemical equation, b is coefficient of reactant 2 from the balanced chemical equation, c is coefficient of product 1 from the balanced chemical equation, d is coefficient of product 2 from the balanced chemical equation, and S_f is entropy of formation

Change in entropy of formation of water given in Eq. (8.28) and (8.29) of a chemical reaction can be determined using the above-stated procedure. Eq. (8.28) can be redrawn for Eq. (8.30) where water is the product, and hydrogen and oxygen are the reactants.

$$\Delta S = (S_f)_{H_2} + \frac{1}{2}(S_f)_{O_2} - (S_f)_{H_2O} \tag{8.30}$$

$$\Delta S = 0.13066 + \frac{1}{2}(0.20517) - 0.06996$$

$$\Delta S = 0.163\text{kJ/mol}$$

Similarly, the Gibbs free energy of the reaction is determined as follows:

$$\Delta G = \Delta H - \text{T} \cdot \Delta S$$

$$\Delta G = 286\text{kJ/mol} - 298(0.163\text{kJ/mol})$$

$$\Delta G = 237.4\text{kJ/mol}$$

The positive Gibbs free energy of the reaction indicates that the reaction is not spontaneous. So to complete the reaction, energy is to be supplied from outside. We provide this energy in the form of electrical energy. This electrical energy is provided by applying a potential difference across the electrodes. This potential can be found by the following relation of Eq. (8.31).

$$V_{rev} = \frac{\Delta G}{nF} \tag{8.31}$$

where V_{rev} is reversible potential, ΔG is Gibbs free energy, n is no. of electrons involved in the reaction, and F is Faraday's number (96485C/mol)

$$V_{rev} = \frac{237.4\text{kJ/mol}}{2(96485C/\text{mol})}$$

$$V_{rev} = 1.229 \text{ V}$$

This is the minimum voltage required to carry the electrolysis process at a standard temperature of 25°C. As the temperature is increased, we require less voltage for the electrolysis. The reaction that took place under the above-stated conditions is an endothermic reaction and the process must take energy from the surroundings to carry on at 1.229 V. If the electrolysis is carried at the potential greater than the reversible potential, losses occur in the cell and the process becomes exothermic releasing heat to the atmosphere.

8.5.3.2 Thermolysis

Another mechanism to perform the water-splitting process to generate the hydrogen is the use of thermal energy, either directly or from the chemical reactions. The reaction of the thermolysis is the following.

$$\mathrm{H}_2O + thermal\ energy\ energy \Rightarrow \mathrm{H}_2 + \frac{1}{2}O_2$$

Historically, the heat to perform the chemical reaction of water-splitting was taken from nuclear reaction. With the advancement in concentrated solar power, the future of hydrogen production seems to be renewable energy based. Fig. 8.11 shows the conversion of water into hydrogen and oxygen by solar thermolysis. Solar towers and solar dishes are used for the concentration of the sunlight at the receiver point where the water is split into hydrogen and oxygen.

8.5.3.3 Photolysis

The dissociation of water into hydrogen and oxygen in a series of biological activities in the presence of sunlight is called photolysis. The overall reaction of the photolysis is shown in the following equation.

$$\mathrm{H}_2O + light\ energy \Rightarrow \mathrm{H}_2 + \frac{1}{2}O_2$$

Photolysis of water by solar radiation for the production of hydrogen is an environment-friendly and renewable energy process.

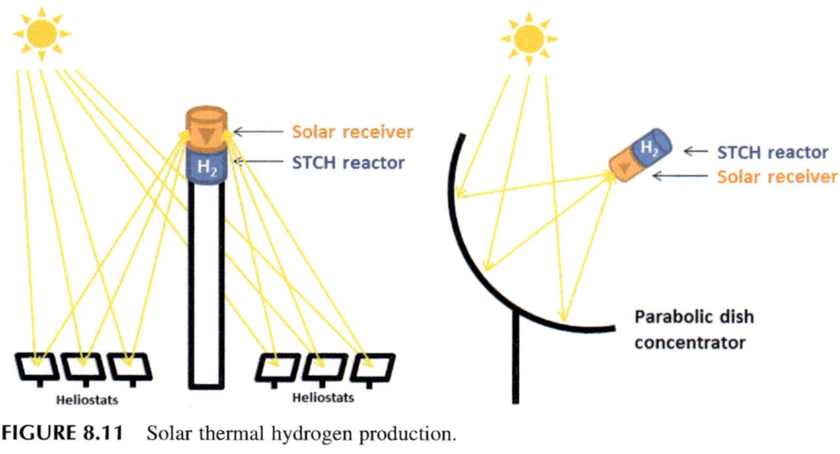

FIGURE 8.11 Solar thermal hydrogen production.

8.6 Economic considerations

The energy required to produce the biofuels must be considered while discussing the economics of bioenergy. For example, the cultivation and harvesting of corn and the production of ethanol from corn uses energy from fossil fuels in producing agricultural equipment, manufacturing the fertilizers, transportation and processing of corn, and the distillation. On the other hand, the energy from the biodiesel is the minor energy gain.

8.7 Conclusion

Biomass is a renewable energy source that is abundantly available on Earth. It could be either woody biomass or agricultural waste. Biomass is categorized as vegetable oil, sugar and starch, lignocellulosic material, and wet biomass. These different types of biomass can be converted into biodiesel, biogas, chemicals, electricity, heat, and hydrocarbons by using different technologies, for example, transesterification, gasification, pyrolysis, anaerobic digestion, palletization, hydrolysis, and fermentation. In rural areas, biomass fulfills the cooking, hot water, and space heating needs.

Problems

Problems 1–14 contain three/four answer options A, B, C, and D. Choose the correct answer.

1. Which of the following is the equation of photosynthesis?
   A. $6CO_2 + 6H_2O \rightarrow 2C_6\mathrm{H}_{12}O_6 + 6O_2$
   B. $C_6\mathrm{H}_{10}O_5 + H_2O \rightarrow C_6\mathrm{H}_{12}O_6 + H_2$
   C. $C_6\mathrm{H}_{12}O_6 \rightarrow 2C_2H_5\,OH + 2CO_2$
   D. $C_6\mathrm{H}_{12}O_6 + 2H_2 \leftrightarrow C_2H_5\,COOH + 2H_2O$
2. Which of the following is not a part of the anaerobic digestion?
   A. Acidogenesis
   B. Acetogenesis
   C. Methanogenesis
   D. Biogenesis
3. The final output of the anaerobic digestion process is
   A. CO_2 *and* CH_4
   B. CO_2 *and* H_2O
   C. CH_4 *and* H_2O
   D. All of the above
4. Which of the following process is used for the production of biodiesel?
   A. Photosynthesis
   B. Transesterification
   C. Pyrolysis
   D. Fermentation

5. Lignocellulosic biomass can be converted into ethanol through which process?
   A. Hydrolysis–fermentation process
   B. Pyrolysis
   C. Gasification
   D. Palletization
6. The conversion process of vegetable oil into biodiesel is known as
   A. Photosynthesis
   B. Transesterification
   C. Pyrolysis
   D. Fermentation
7. Which of the following is a substrate for biogas production?
   A. Municipal solid waste and residential waste
   B. Metallic waste
   C. Gaseous waste
   D. All of the above
8. The proximate analysis determines which of the following?
   A. Ash contents and fixed carbon
   B. Volatile matter and moisture contents
   C. Carbon, hydrogen, and oxygen
   D. A & B
9. Which of the following best indicates the steps of anaerobic digestion?
   A. Waste water feed → biogas → digester → biogas storage → generator
   B. Waste water feed → biogas storage → generator → biogas
   C. Generator → waste water feed → digester → biogas → biogas storage
   D. Waste water feed → digester → biogas → biogas storage → generator
10. Which of the following is not a thermochemical process to produce hydrogen?
    A. Pyrolysis
    B. Dark fermentation
    C. Photofermentation
    D. Biophotolysis
11. Which of the following is produced in acetogenesis?
    A. Ethanol
    B. Acetate
    C. Acetone
    D. Ketone
12. Incomplete combustion of biomass in the presence of an oxidizing agent is called
    A. Gasification
    B. Pyrolysis
    C. Biophotosynthesis
    D. Fermentation

13. Hydrogen can be produced by the process
   A. Thermochemical process
   B. Biological process
   C. Water-splitting
   D. All of the above
14. Which of the following is the cold weather characteristic of a fuel?
   A. Cloud point
   B. Boiling point
   C. Flashpoint
   D. Melting point
15. Enlist the feedstock materials that can be used as a substrate in the biogas digester.
16. Write down the three biological processes that are used to produce hydrogen.
17. Differentiate between pyrolysis and fermentation.
18. What is producer gas and how it is formed?
19. What is pellet fuel?
20. Differentiate between proximate analysis and ultimate analysis.
21. Differentiate between electrolysis and water-splitting.

References

[1] C. Bonechi, M. Consumi, A. Donati, G. Leone, A. Magnani, G. Tamasi, C. Rossi, 1—Biomass: an overview, in: F. Dalena, A. Basile, C. B. T.-B. S. for the, F. Rossi (Eds.), Bioenergy Systems for the Future, Woodhead Publishing, 2017, pp. 3–42.

[2] D.Y. Murzin, I.L. Simakova, 7.21—Catalysis in biomass processing, in: J. Reedijk, K. B. T.-C. I. C. I. I.E. (SecondPoeppelmeier) (Eds.), Comprehensive Inorganic Chemistry II, Elsevier, Amsterdam, 2013, pp. 559–586.

[3] R.A. Labatut, J.L. Pronto, Chapter 4—Sustainable waste-to-energy technologies: anaerobic digestion, in: T.A. Trabold, C. W. B. T.-S. F. W.-T. S. Babbitt (Eds.), Sustainable Food Waste-to-Energy Systems, Academic Press, 2018, pp. 47–67.

[4] M. Kamran, Current status and future success of renewable energy in Pakistan, Renew. Sustain. Energy Rev. 82 (2018) 609–617. Available from: https://doi.org/10.1016/J.RSER.2017.09.049.

[5] R. Luque, J.A. Melero, 1—Introduction to advanced biodiesel production, in: R. Luque, J. A. B.B. P. T.-A.Melero (Eds.), Advances in Biodiesel Production, Woodhead Publishing, 2012, pp. 1–9.

[6] F. Ma, M.A. Hanna, Biodiesel production: a review1Journal Series #12109, Agricultural Research Division, Institute of Agriculture and Natural Resources, University of Nebraska–Lincoln.1, Bioresour. Technol. 70 (1) (1999) 1–15, https://doi.org/10.1016/S0960-8524(99)00025-5>.

[7] H. Jiang, Y. Wu, H. Fan, J. Ji, Hydrogen production from biomass pyrolysis in molten alkali, AASRI Proc 3 (2012) 217–223. Available from: https://doi.org/10.1016/j.aasri.2012.11.036.

Chapter 9

Geothermal energy

Muhammad Rayyan Fazal and Muhammad Kamran
Department of Electrical Engineering and Technology, Riphah International University, Pakistan

9.1 Introduction

Geothermal energy is natural energy in the form of heat that is produced and preserved within the Earth. The term "geothermal energy" is taken from the Greek words, "geo" and "thermo"—earth and heat respectively—whereas "energos" means active. Hence, the overall term may be referred to as the movement of active heat energy within the Earth. The temperature of the Earth increases with its depth and this geothermal gradient allows continuous heat conduction from the Earth's core to the surface. This is also called "terrestrial heat-flow." If the heat flow rate is high and energy extraction is economical, then humans can utilize this heat for multiple purposes, that is, electricity generation, heating, and agriculture use. The average heat flow to the surface of the Earth is 50 mW/m^2. The temperature increases 25°C–30°C with every kilometer of the Earth's depth, with some regions having this gradient as high as 150°C. It is the only source of energy that can be used to supply baseload or dispatchable power without weather constraints. Due to the increasing demand for switching to clean energy production, the need for research and development in engineered geothermal systems (EGS) is becoming critical. Recently the development in geothermal systems has enabled the elimination of the need for natural hydrothermal resources. It can be done by fracturing the subsurface rocks and injecting water. The water is heated when it comes in contact with the rocks, and production wells are then utilized to send it to the surface.

Geothermal energy is considered renewable if it can supply according to demand. Its high energy density and continuous availability make it highly desirable and give it an advantage over solar and wind. However, it has its limitations, like transportation constraints, unlike fossil fuels. It can be used directly to extract any advantage out of it. The distribution of geothermal energy resources follows an irregular pattern due to the unequal distribution of water geysers, hot springs, fumaroles, silica sinter terrace, alga mat, lava flows, hot pools and lakes, steaming pools, and volcanoes. These resources

Renewable Energy Conversion Systems. DOI: https://doi.org/10.1016/B978-0-12-823538-6.00006-3

are highly associated with geological and hydrological patterns. The rich resources of heat energy are considered to be the areas located near volcanic chains or rings of fire where tectonic plates meet and frequent volcanic activities can be observed. The potential target is explored through a comprehensive analysis followed by geochemical studies and the drilling process. These complex studies are followed by geophysical surveys.

Mainly there are two types of geothermal systems, that is, vapor-dominated and liquid-dominated. Vapor-dominated systems are very rare and uncommon with water and vapors coexisting in the reservoir. Geysers (California, United States), Kamojang, Darajat (Java, Indonesia), Larderello (Italy), and Matsukawa (Japan) are found in the literature confirming the areas with required pressure of vapor-dominated fields [1]. The majority of the geothermal systems belong to the liquid-dominated type. These types of systems have reservoirs with an average temperature of 200°C capable of producing electrical power in the range of 2–10 MW. They are generally found in active tectonic environments having a moderate to high temperature range. Hence, most of them can be found near volcanic chains. The rocks in these systems are naturally shaped to support the liquid flow.

The updated data of the world geothermal congress held in 2020 provide a deep insight regarding the current trends of geothermal energy and future trends. A very comprehensive country-wise data collection (2015–2020) and forecasting are available in the literature [2]. During the period of 2015–2020, a 3.649 GW (27%) increase in geothermal energy was observed. The global trend of geothermal capacity increase with 5-year time intervals (2010–2020) is provided in Fig. 9.1, with the next 5 years forecast. Five new countries have for the first time started energy production through geothermal means: Chile (48 MWe), Honduras (35 MWe), Croatia (16.5 MWe),

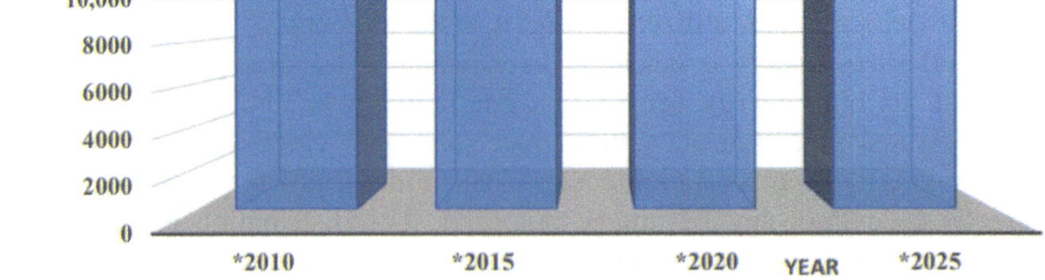

FIGURE 9.1 Global trend of installed geothermal capacity till 2020- and 5-year forecasting (2020–2025) [2].

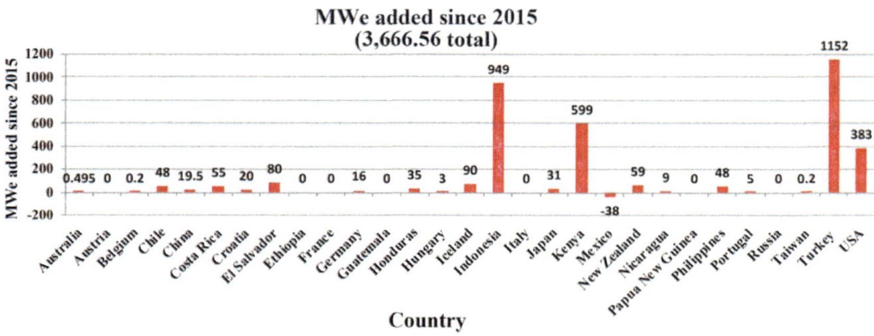

FIGURE 9.2 Country-wise MWe capacity addition during the period 2015–2020 [2].

Hungary (3 MWe), and Belgium (0.8 MWe). In these 5 years, the statistics of the global geothermal industry are 1159 new wells have been drilled, $10.367 million have been spent on power projects, 30,491 person-years have been allocated for the power projects, and 3666.56 MWe have been added. The country-wise breakdown of capacity addition is given in Fig. 9.2. The countries with the greatest total installed capacity are the United States, Indonesia, the Philippines, and Turkey. Indonesia alone has the world's four largest capacity plants, with Gunung Salak, the largest, having 375 MWe capacity. Some of the nations are planning to add geothermal power capacity into their systems in near future. These countries include Caribbean nations, Canada, Greece, Iran, and Ecuador [2].

A comprehensive review based upon the current data and future predictions up to 2050 provides deeper insight into these systems [3]. The potential for the global power capacity of geothermal energy systems is predicted to be 6 TWe for the year 2030 and 108 TWe for the year 2050. It reaches 200 TWe if constraints such as water stress and cost stress are excluded. The top five countries with the greatest potential by 2050 are Russia, China, Brazil, the United States, and Canada. Also, it is predicted that implementing the full potential of enhanced geothermal energy systems by 2050 can provide five times the overall energy demand of 2016. Keeping resource quantification methods aside, the amount of geothermal energy extraction will be influenced by the government policies and strategies as well as adaptation of advanced technologies [3].

9.2 Geothermal resources

Geothermal resources are part of the geothermal field heat energy and can be collected by fluid; it can be steam or hot water or a blend of both. The two main categories of geothermal resources are natural and artificial. The natural resources have their heat creation and supply mechanism, where artificial resources are created through digging wells, installing pumps, heat exchange systems, and connecting pipes. Wells are drilled into the natural resources where direct heat extraction is not possible, hence bringing the hot water or steam to the upper surface by indirect or artificial means. This can be then utilized in geothermal power plants for electricity production. The most crucial part related to both of these systems is heat transfer. Most of the classification of geothermal resources is based upon the heat transfer mechanism that happens between the fluid and rocks.

The three main geothermal resources according to the heat-producing sources are hot water reservoir, dry rock, and magma. There are many other categories based on the heat of crust. The different available geothermal resources are:

- *Hydrothermal resources*: these are also regarded as hot water or steam reservoir. The hot water reaches the Earth's surface but remains inside the Earth confined in porous rocks from natural hot water or steam resources. It happens because the surface water travels through the fractures and faults of the rocks deep into the hot region of the Earth's crust. It is the natural phenomenon of the water to travel back to the surface upon heating due to its lower density and convention process. This can occur in the form of mud holes, springs, fumaroles, and geysers. If this rising water is encountered with a very high temperature, it converts into steam. The main location of these resources is seismic areas along with tectonic plates. The rising water facing impermeable blockage is collected underground with temperature much higher than that reaching the surface, that is, it can reach as high as 350°C. It can be utilized through drilling and bringing hot water and steam to the surface. Most of the reservoirs are hot springs, there are some steam producing reserves but these are very rarely available. The current research to gather the maximum benefit out of these reserves is underway. However, different researchers have presented the analysis and provided feedback to make it more effective [3–5].
- *Hot dry rock*: these types of reservoirs are located deep down with no access to water. They comprise very hot and dry rocks but heat can only be recovered by artificial means, that is, injecting the fluid into these reserves. Hence, to bring the available heat to the Earth's surface, fluid is injected by drilling into the rocks, creating an artificial reservoir, and then extracted from the well for practical utilization. It is estimated that if dry rock techniques are exploited in a better way then 5–10 times more geothermal electricity production is possible. The design and evaluation of the fracturing projects related to hot dry rocks can be found in recent studies [6].
- *Geopressured resources*: fossil fuel-rich areas containing a large quantity of methane are often encountered with the waters underground. They are mixed with water and trapped in sand layers. They often have high

pressure and the temperatures reaching 200°C. These areas are suitable in the sense that both water and methane can be extracted to produce electricity. Large reservoirs can produce the required heat energy for quite a long time. However, there are some associated disadvantages in the context of geothermal energy systems. The overall framework regarding their practical impact, controlling, and prevention mechanism is available in the literature [7].

- *Magma*: this is a semiliquid rock found beneath the Earth's surface. It is extremely hot having different types with an overall temperature range of 650°C-1200°C. This enormous heat energy at the same time requires very special arrangements to make it available for heat collection and electricity generation. This is an exciting research area due to the availability of continuous and huge amounts of heat. Research activities are trying to present the possible advantages and the ways to extract the heat safely [8,9].
- *Low temperature-based resources*: the places where temperatures are below 150°C are considered to be the low-temperature resources. They can be used to generate electricity and various other purposes. The research suggesting their availability, usefulness, and practical applications in different aspects is currently in progress [10–12].

The geothermal resource assessment and estimating the reserve capacity is critical for the practical implementation of geothermal energy power plants. There are multiple ways to achieve the goal to decide the feasibility of the project in light of features such as potential capacity, finance, and operational requirements. The two most suitable approaches for geothermal resource assessment include volumetric methods and reservoir modeling. Volumetric methods are used for resource estimation at the initial stages. Reservoir modeling is performed for the operational and management of relevant requirements in terms of sustainability. There are some techniques offering advantages over others but the main approach remains the same [13].

9.3 Geothermal energy conversion mechanism

Electricity production from geothermal resources is now considered a well-established and mature technology. Geothermal electric energy is gaining popularity mainly because of its high load factor, that is, capable of producing more energy per megawatt per year as compared to similar capacity solar and wind power systems. The energy conversion mechanism is mainly made up of the heat extraction mechanism and the energy production unit. There are three main technologies used to convert geothermal energy into electricity. The first is *open-loop systems*, using geothermal fluids directly integrated with steam turbines to drive generators, for example, dry steam plants and single- or double-flash plant. The second technology is *closed-loop systems*, where the heat extraction from the geothermal fluid is done and then transferred to turbines to drive generators through the heat exchange process. They are regarded as closed-loop systems because the condensed fluid is sent back to the generator. They are used for hydrothermal and low-temperature resources. The third technology is called a combined, mixed, or hybrid system, combining open and closed-loop systems. The levelized cost of electricity related to the geothermal based power plants is \$0.04–0.14 per kWh. The overall 25 years of the economic cycle is assumed with annual maintenance expenses of \$110 kWh [14].

9.3.1 Dry steam power plants

Dry or direct stream power plants are extremely rare due to the limited availability of dry steam resources. They require a temperature greater than 150°C and require 99.99% dry steam [14]. They were the first type of geothermal power plant, first established in 1904 in Laradarello, Italy [15]. If any resource having the required temperature and pressure is found, the overall process becomes quite simple, economical, and efficient. Steam is collected through multiple wells and purified from mist and particles before feeding into the turbine. Mist separators and particle removers are used for the purification process. The turbine is meant to receive steam as input to rotate the shaft of the turbine and this mechanical energy is converted to electrical energy utilizing a generator. The condenser collects the cold steam which is then pumped back through the injection well. The overall process is shown in Fig. 9.3. The emission of excessive steam leaves 15%–20% of the condensed geofluid available for reinjection, whereas releasing gas into the atmosphere and condensed water on the surface is considered a more economical option. However, the continuous utilization of geothermal fluid without replacement may degrade the fluid quality and associated properties. The size of dry steam power plants ranges between 8 and 140 MW [14].

9.3.2 Flash steam power plants

These types are the most common types of geothermal electricity plants where steam is separated from the water through a separation process called flashing. The steam is then directed to turbines in the same way as in the dry steam power plants described before. Flash power plants use reservoirs with a temperature greater than 180°C. Most of the geothermal resources contain a fluid that is a blend of steam and liquid. The easiest way is to separate both and use steam as input to turbines but steam is in very low quantity if used without any further processing. So, a special arrangement is made to take further benefit of this mixture. The fluid is passed through a flash separator where the pressure is kept low to convert the hot liquid into saturated

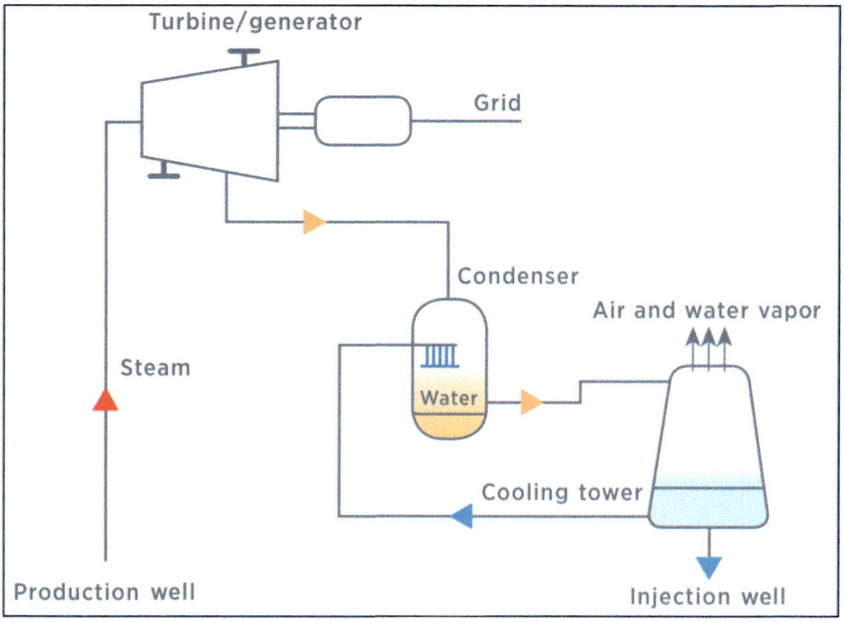

FIGURE 9.3 Dry steam power plant [16].

vapors due to a sudden pressure drop. The steam is then separated and fed to the turbines and liquid is pumped back to the geothermal reservoir. In the turbine, the steam expands and gets cooled while turning the turbine shafts, hence producing mechanical energy. The shaft is connected to the generator to produce electric power.

Flash power plants are further categorized into single, double, and triple-flash plants. The single-flash plants have the disadvantage of not utilizing most of the geothermal energy. Double- and triple-flash systems are employed to overcome this deficiency. The fluid separated from the steam, remaining hot, and contains a high amount of energy. This fluid is fed to the high-pressure and low-pressure turbines as shown in the double-flash systems in Fig. 9.4. The high-pressure steam in the first cycle is provided to the steam turbine and liquid is fed to the second stage. The second stage collects the steam and feeds the liquid to the low-pressure chamber of the turbine (a double-flash mechanism). The liquid is then checked for the energy content and can be used for the third-stage steam separation if there is a high energy content (a triple-flash mechanism), or otherwise pumped back to the injection well to complete the cycle. The double-flash plants are capable of producing 25% more energy as compared to the single-flash plants and can return nearly 85% of the liquid to the reservoirs. The power production capacity of single-flash power plants is 0.2–80 MW, double-flash plants

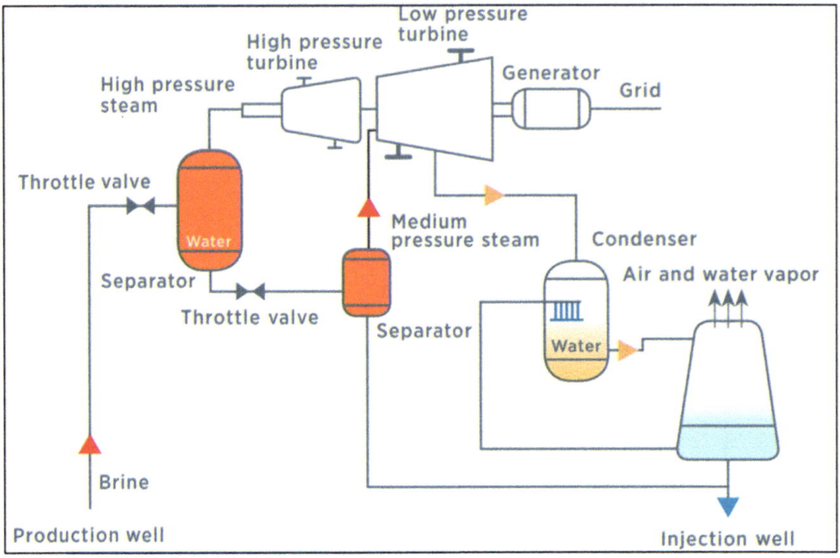

FIGURE 9.4 Flash steam power plant [16].

have 2–110 MW capacity and capacity for triple-flash plants is 60–150 MW [14].

9.3.3 Binary cycle power plants

Binary cycle plants are employed for lower temperature reservoirs as compared to dry steam and flash steam plants which are efficient in the range of 180°C-350°C. For lower temperatures, that is, 107°C-180°C, binary cycle power plants can be employed. It is possible to work on temperatures even lower than that 100°C but efficiency decreases. As the name suggests, it has two cycles, where the first cycle collects the heat from the geothermal resource-based fluid, and in a second cycle, it is used to boil a second fluid having a lower boiling point. This process is accomplished through a heat exchanger known as an evaporator. The second fluid in the heat exchanger provides the steam which is then fed to the turbine for further action. the steam is passed through the condenser to make it liquid again, hence closing the second cycle.

Binary plants use the organic Rankine and Kalina cycle. These power plants offer the advantage of using separate geothermal and secondary fluids, allowing electricity production through low-temperature geothermal resources. Also, they pump 100% geothermal fluid back to the injection well, hence offering the least environmental impact with little or no emissions. On average, they are lower capacity plants with a power-producing capacity of 1–50 MW [14] (Fig. 9.5).

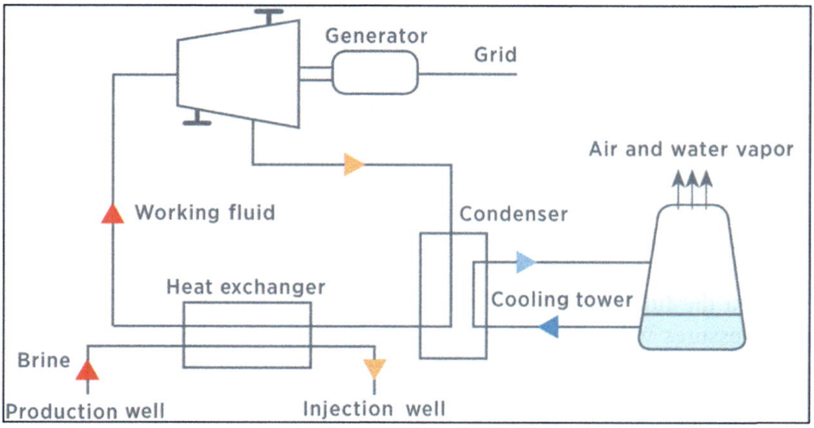

FIGURE 9.5 Binary cycle plant [16].

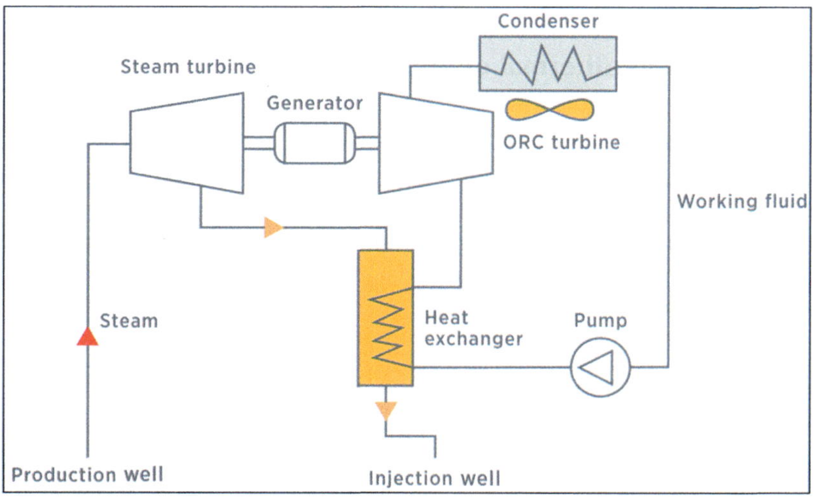

FIGURE 9.6 Geothermal based combined cycle plant [16].

9.3.4 Geothermal combined cycle power plants

Combined cycle plants are also known as mixed or hybrid geothermal power plants. The binary cycle plants can be used in combination with the flash plants to produce a higher amount of electricity, as shown in Fig. 9.6. This is an efficient way for geothermal fluid utilization but highly dependent upon resource availability. The flash plant extracts the steam and the binary plant utilizes hot residual liquid, but both of these plants work independently. Hence, adding the Rankine cycle produces extra electricity from the extra liquid which would become waste otherwise. The basics are the same as those of simple geothermal power plants with the difference of combining two heat sources. Steam turbines use flash tanks directly, and the binary cycle uses drained hot water employing the organic Rankine cycle and Kalina cycles. These hybrid power plants provide higher power output in the range of a few MW to 10 MWe [14].

There are several places where geothermal plants can be employed to extract energy. Advancement of the technology can make allow it to extract maximum power out of hard dry rock-based geothermal resources. The undersea located geothermal resources could also be exploited for their utilization in the future. Some of the reservoirs have very high temperatures and high pressures with supercritical waters capable of producing 10 times more energy than a sample reservoir. However, future technology might enable us to find ways to extract the enormous heat from magma-based geothermal resources. There is plenty of room for future research and development, and the identification and drilling process are associated with some issues. These issues are identified and can be analyzed to suggest future directions.

9.4 Use of geothermal energy

The use of geothermal energy can be found in several applications nowadays. As there is a huge growth observed in its usage. Low-temperature reservoirs can be used for direct heat extraction including house heating, bathing, agriculture use, fish farming, etc. It can be indirectly used to convert heat into electrical energy through thermodynamic cycles employing turbines and generators. The former is regarded as direct use, whereas the latter is indirect use of geothermal energy.

9.4.1 Indirect uses of geothermal energy

The electricity production methods discussed above are examples of indirect use of geothermal energy. Where heat is not directly used for useful purposes, it is used indirectly, that is, to turn the turbine shaft, hence converting heat energy to mechanical energy. Which is then converted to electric energy employing any generator. It can be then transmitted through transmission lines over longer distances.

9.4.2 Direct uses of geothermal energy

Direct use of geothermal energy involves the direct usage of heat for multiple purposes. It is the most common and oldest way. It is especially useful for cold countries. In previous years there has been a sharp increase observed in the direct use of geothermal electricity, as shown in Fig. 9.7. The total installed capacity in 2019 was observed to be 107,727 MWt, with 8.7% per annum growth and 52% overall growth as compared to the World

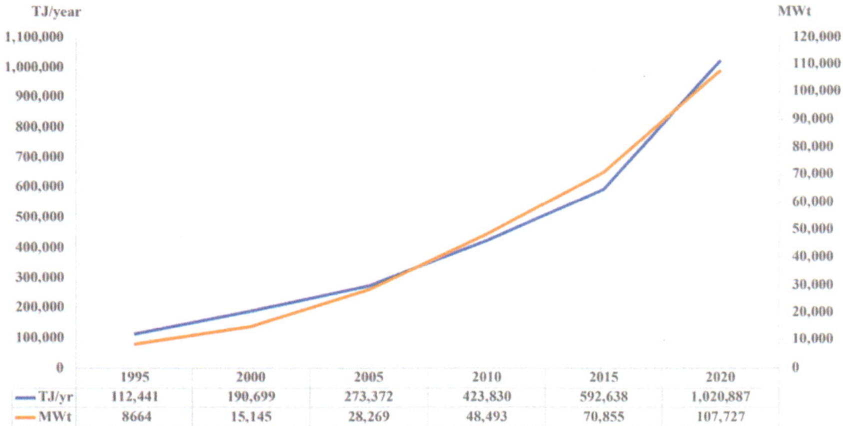

FIGURE 9.7 Installed capacity and annual utilization of directly used geothermal energy [17].

Geothermal Congress 2015 (WGC2015). The overall energy consumption in 2019 was found to be 1,020,887 TJ, with 11.5% annual growth and a 72.3% increase since the world geothermal congress 2015 (WGC2015). Continent-wise Asia is on top followed by Europe and the Americas [17].

There are many applications of direct use of geothermal energy.

- *Space heating*: it is the simplest method to use geothermal energy by directing the steam or geothermal liquid to space or any object. This heat can be used directly or used to heat any other medium such as water or air which can be used to heat homes. It includes stand-alone space heating or district heating. Space heating systems have increased 83.8% in annual energy usage (162,979 TJ/year) and 68% in total installed capacity (12,768 MWt) [17]. China, Iceland, Turkey, France, and Germany are leaders in space heating systems making 90% of the total district heating usage [17].
- *Heat pumps*: these pumps work on the reverse refrigeration phenomenon. They take heat energy from a cold source and release it into a hot sink. Evaporators can be placed in hot sinks and condensers can be placed in cold areas to convert an air-conditioning system into a heat pump. They are employed to extract the outside heat and release it inside the houses. But they have to work harder in cold regions and winter seasons due to lower in/out temperatures. Hence, this creates a need for getting the required heat from under the Earth's surface, which maintains its inner temperature irrespective of the outside temperature. They are in the largest use globally with 71.6% of overall installed capacity (77,547 MWt) and 59.2% energy usage (599,981 TJ/year). most of them are located in North America, Europe, and China. Individual unit size ranges between 5.5 (residential) and 150 kW (commercial).

- *Covered ground heating (greenhouse)*: greenhouses can use geothermal resources for multiple purposes, including growing vegetables, fruits flowers, tree seedlings, and fish farming. Greece and Iceland have reported having covered ground heating. The major cost is associated with the labor, providing an advantage to developing nations over developed countries. An increase has been observed of 24% in installed capacity (2459 MWt) and a 23% increase was seen in energy usage (35,826 TJ/year). At WGC 2020, 32 countries reported being active users of geothermal greenhouse heating, with Turkey, China, Netherland, Russia, and Hungary as the leading countries, making up 83% of the total world's geothermal energy used in greenhouses and covered grounds [17].
- *Aquaculture*: the farming of marine species for various uses can be developed by extracting the necessary heat from geothermal resources. Salmon, tilapia, trout, and bass are found to be the most common breeds along with alligators, shrimps, prawns, and lobsters. Using the covered strategy for pools reduces the energy requirements to one half, however they are very uncommon. Aquaculture has observed increases of 36.5% in installed capacity (950 MWt) and 13.5% energy usage (13,573 TJ/year). China, the United States, Iceland, Italy, and Israel are among the top 21 countries that have reported this usage type, with an overall 92% combined contribution [17].
- *Crop drying*: the heat from geothermal resources is also used for crop drying purposes. Various grains, vegetables, and fruits are dried including onions, wheat, cereals, fruits, coconut meat, timber, and seaweed. China, France Hungary, the United States, and Japan are the largest users of countries that have reported this form of usage. Crop drying has witnessed an increase of 59.6% in installed capacity (257 MWt) and a 73.8% increase in energy usage (3529 TJ/year) [17].
- *Miscellaneous proposes*: Geothermal energy can have direct use for bathing and swimming purposes as reported by 53 countries. It is used in industrial process heat applications, as reported by 14 countries, with China on top. This application requires high energy consumption due to the huge volume of industries. Examples of this application type are milk pasteurization, chemical extraction, leather industry, carbonated drinks, water bottling, CO_2 extraction, paper processing, salt and iodine extractions, boric acid production, etc. Other applications include snow melting and space cooling, animal farming, desalination of water, and sterilization of bottles [17].

9.5 Environmental effects

Geothermal electricity is regarded as green and renewable but some emissions are associated with the nature and type of plants and resources.

Methane, CO_2, and hydrogen sulfide are the main contributors in this regard. However, methane, sulfur dioxide, ammonia, and hydrogen are also found in small quantities. These emissions seem acceptable if compared with fossil fuel emissions. It has a global average CO_2 emission factor 122 g/kWh of 122 g/kWh, which varies from 34 g/kWh (Iceland power plants) to 330 g/kWh (Italian power plants). Turkey however has a high CO_2 emission, even higher than the fossil fuel emissions [18]. Quantitative assessment of environmental risks related to geothermal energy can be found in the literature [19,20].

Geothermal is considered to be green and the cleanest form of energy as it significantly reduces greenhouse gas emissions, acid rain, and other air pollution forms. It has major disadvantages, including higher capital cost, silica scaling, being subject to geological changes, hydrothermal eruption risks, unavailability of standardized data, the uncertainty of resource availability, and low efficiency. Its advantages include the lower land-related requirement, low water usage, CO_2 control, continuous availability irrespective of the climatic conditions, high reliability, and local job creation [21]. In a nutshell, it can be proven to be a potential alternative against conventional electricity generation technologies. There is a huge resource base for geothermal energy and it is well distributed throughout the globe. Efforts are now being made to access offshore geothermal resources at a pace never seen before [22].

Example 9.1

A geothermal power plant takes saturated working fluid at a temperature of 150°C from the production well at a rate of 230 kg/s and sends the working fluid back to the injection well at a temperature of 25°C. Find the following if the turbine is producing output power of 21.00 MW.

(a) The input heat from the production well to the power plant
(b) Actual thermal efficiency
(c) Maximum thermal efficiency
(d) The heat rejected

Given data

$$T_{production\ well} = 150^{\circ}\text{C}$$

$$T_{injection\ well} = 25^{\circ}\text{C}$$

$$\dot{m} = 230\ \text{kg/s}$$

$$W_{out} = 21\ \text{MW}$$

Solution

Using saturated steam properties we consult Table A-4 to know the enthalpies at the given temperature states.

$$T_{production\ well} = 150°C \Longrightarrow h_{production\ well} = 632.18\ kJ/kg$$

$$T_{injection\ well} = 25°C \Longrightarrow h_{injection\ well} = 104.83\ kJ/kg$$

(a) The input heat from the production well to the power plant can be determined by taking the difference of the enthalpies of fluid at the production well and the injection well using the following equation.

$$Q_{in} = \dot{m}(h_{production\ well} - h_{injection\ well})$$

$$Q_{in} = 230\ kg/s(632.18\ kJ/kg - 104.83\ kJ/kg)$$

$$Q_{in} = 121,290\ kW$$

(b) Actual thermal efficiency can be determined as following

$$\eta_{th} = \frac{W_{out}}{Q_{in}} = \frac{21\ MW}{121.290\ MW} = 17.3\%$$

(c) Maximum thermal efficiency can be determined by the following expression of temperatures.

$$\eta_{th} = \left(1 - \frac{T_{injection\ well}}{T_{production\ well}}\right)$$

$$\eta_{th} = \left(1 - \frac{25 + 273}{150 + 273}\right)$$

$$\eta_{th} = 29.5\%$$

(d) The rate of heat rejection is calculated by the following equation

$$Q_{out} = Q_i - W_{out}$$

$$Q_{out} = 121.290\ MW - 21\ MW$$

$$Q_{out} = 100.29\ MW$$

Problems

Problems 1–10 contains four answer options A, B, C, and D. Choose the correct answer.

1. The process of producing energy by utilizing heat trapped inside the Earth's surface is called _________.

A. Hydrothermal energy
B. Geothermal energy
C. Solar energy
D. Wave energy

2. What is hot molten rock called?
   A. Lava
   B. Magma
   C. Igneous rock
   D. Volcano
3. How many kinds of geothermal steams are there?
   A. 2
   B. 3
   C. 4
   D. 5
4. What does EGS stand for in geothermal energy?
   A. Engraved geothermal systems
   B. Enhanced geothermal system
   C. Exhaust gas system
   D. Engineered geophysical system
5. Who invented first geothermal plant?
   A. Michael Faraday
   B. Piero Ginori Conti
   C. Enrico Fermi
   D. Guglielmo Marconi
6. A spring that shoots jets of hot water and steam into the air is called a ________
   A. Mine hole
   B. Geyser
   C. Hot spring
   D. Mud pot
7. Which is the most common type of geothermal plant?
   A. Dry steam
   B. Flash
   C. Binary
   D. Wet steam
8. What is the efficiency of a geothermal plant?
   A. 28%
   B. 15%
   C. 42%
   D. 30%
9. Direct conversion of heat energy into electrical energy is possible through
   A. MHD generators
   B. Fuel or solar cells

C. Thermionic converters.
D. Both A and C

10. In geothermal power plants wastewater is
    A. Discharged back into the Earth
    B. Discharged into the sea
    C. Recirculated after cooling in cooling towers
    D. Evaporated in ponds
11. What is a flash steam power plant?
12. Explain the dry steam power plant?
13. What are the cycles that are involved in the combined cycle. State the flow process of the geothermal combined cycle power plant.
14. Briefly explain the direct and indirect uses of geothermal energy.
15. Is geothermal energy environment-friendly? Explain your answer.
16. What is the difference between open loop and closed loop geothermal power plants.

References

[1] R. Allis, Insights on the formation of vapor-dominated geothermal systems, in: Presented at the Proceedings of World Geothermal Congress, 28 May–10 June, 2000. <https://www.geothermal-energy.org/pdf/IGAstandard/WGC/2000/R0610.PDF>

[2] G.W. Huttrer, Geothermal power generation in the world 2015–2020 update report, in: Proc. World Geothermal Congress 2020, Reykjavik, Iceland, 2020.

[3] A. Aghahosseini, C. Breyer, From hot rock to useful energy: a global estimate of enhanced geothermal systems potential, Appl. Energ. 279 (2020) 115769.

[4] Z. Lyu, X. Song, G. Li, Numerical analysis of characteristics of reaction in hydrothermal jet drilling for geothermal energy, Geothermics 77 (2019) 62–74.

[5] X. Bu, K. Jiang, Y. He, Performance analysis of shallow depth hydrothermal enhanced geothermal system for electricity generation, Geothermics 86 (2020) 101847.

[6] Z. Zhou, et al., Investigation on fracture creation in hot dry rock geothermal formations of China during hydraulic fracturing, Renew. Energ. 153 (2020) 301–313.

[7] B. Yuan, D.A. Wood, A holistic review of geosystem damage during unconventional oil, gas and geothermal energy recovery, Fuel 227 (2018) 99–110.

[8] W.A. Elders, G.Ó. Friðleifsson, A. Albertsson, Drilling into magma and the implications of the Iceland Deep Drilling Project (IDDP) for high-temperature geothermal systems worldwide, Geothermics 49 (2014) 111–118.

[9] G. Sosa-Ceballos, J.L. Macías, D.R. Avellán, N. Salazar-Hermenegildo, M.E. Boijseauneau-López, J.D. Pérez-Orozco, The Acoculco Caldera complex magmas: genesis, evolution and relation with the Acoculco geothermal system, J. Volcanol. Geotherm. Res. 358 (2018) 288–306.

[10] A. Atiz, H. Karakilcik, M. Erden, M. Karakilcik, Investigation energy, exergy and electricity production performance of an integrated system based on a low-temperature geothermal resource and solar energy, Energy Conv. Manage. 195 (2019) 798–809.

[11] A. Mauro, J.C. Grossman, Street-heat controlling road temperature via low enthalpy geothermal energy, Appl. Therm. Eng. 110 (2017) 1653–1658.

[12] P.A. Østergaard, H. Lund, A renewable energy system in Frederikshavn using low-temperature geothermal energy for district heating, Appl. Energy 88 (2) (2011) 479–487.

[13] A.E. Ciriaco, S.J. Zarrouk, G. Zakeri, Geothermal resource and reserve assessment methodology: overview, analysis and future directions, Renew. Sustain. Energy Rev. 119 (2020) 109–515.

[14] IRENA, Geothermal power, technology brief. <https://www.irena.org/-/media/Files/IRENA/Agency/Publication/2017/Aug/IRENA_Geothermal_Power_2017.pdf>, 2017.

[15] <https://www.energy.gov/eere/geothermal/electricity-generation>

[16] IRENA, Project navigator technical concept guidelines for geothermal projects 2017. <https://navigator.irena.org/index.html>, 2017.

[17] J.W. Lund, A.N. Toth, Direct utilization of geothermal energy 2020 worldwide review, in: Proceedings of World Geothermal Congress, Reykjavik, Iceland, 2020.

[18] T. Fridriksson, A.M. Merino, A.Y. Orucu, P. Audinet, Greenhouse gas emissions from geothermal power production, in: Proceedings of 42nd Workshop on Geothermal Reservoir Engineering, Stanford University, Stanford, CA, 2017.

[19] S. Chen, Q. Zhang, P. Andrews-Speed, B. McLellan, Quantitative assessment of the environmental risks of geothermal energy: a review, J. Environ. Manage. 276 (2020) 111–287.

[20] D. Moya, J. Paredes, P. Kaparaju, Technical, financial, economic and environmental pre-feasibility study of geothermal power plants by RETScreen—Ecuador's case study, Renew. Sustain. Energy Rev. 92 (2018) 628–637.

[21] A. Anderson, B. Rezaie, Geothermal technology: trends and potential role in a sustainable future, Appl. Energy 248 (2019) 18–34.

[22] M.-C. Suárez-Arriaga, J. Bundschuh, F. Samaniego, Assessment of submarine geothermal resources and development of tools to quantify their energy potentials for environmentally sustainable development, J. Clean. Prod. 83 (2014) 21–32.

Index

Note: Page numbers followed by "*f*" and "*t*" refer to figures and tables, respectively.

T

9780128235386